雅典文化

商務英文
E-m@il

Business E-mail

分段解析e-mail寫作重點，利用簡單、實用的句型，
寫出一封漂亮得體的商務e-mail。

張瑜凌 編著

前　言　提高 e-mail 溝通效率

　　身為職場一份子的你，如果對於利用e-mail作為溝通的工具，還顯得懵懵懂懂，甚至無所適從，就實在是與社會職場脫節了！

　　而如果你深刻瞭解e-mail在商務溝通上所能發揮的力量，便已經具備職場競爭力的實力了！但是您知道該如何寫出一封得體又能幫助您有效溝通的商務e-mail嗎？

　　英文寫作並不是要洋洋灑灑地寫文章，而是能夠利用e-mail做為溝通、聯絡、維繫彼此keep me informed的管道。因此艱澀的單字、複雜的文句都是商務英文e-mail寫作的忌諱。請記住，商務英文e-mail的首要認知是快、狠、準的寫作技巧。

　　何謂快、狠、準的寫作技巧？所的「快」，就是必須利用有效的溝通效率時間，千萬不要拖過溝通的黃金期，也就是儘速完成信件往返，不要讓對方寄給你的信件石沈大海。

　　而所謂的「狠」，就是不拖泥帶水，字字句句都是溝通

的利器，不必花太多時間在無謂的說明。

　　而所謂的「準」，就是針對此 e-mail 的主題，精準地回應對方所提出的問題、需求，千萬不要離題太遠，e-mail 可不是讓你練文筆的工具啊！

　　本書「商務英文 e-mail」從 e-mail 的基礎架構開始，到寫作禮儀、句型分析、商務溝通，逐步地帶領您完成一封商務英文 e-mail 所應該具備的寫作實力。

Part 5 禮貌性用語

商務英文 E-m@il

Part 6 事務性洽談

Part 1 撰寫商務英文 e-mail

　　由於Internet的盛行，成就了e-mail電子郵件的溝通系統，使e-mail成了職場辦公室最基本的配備，因此，除了電話之外，e-mail儼然成為不可或缺的商務溝通工具之一。

　　E-mail郵件除了基本的「雙向溝通」的特點之外，還具有提醒、通知、條列式說明的特色。

　　想要寫好一篇得體又專業的商務e-mail書信，必須具備以下的基本能力：

　　一、通曉商業術語及慣用語

　　二、良好的英文寫作基礎

　　三、得體的書面溝通禮儀

　　本書從e-mail的基本結構開始，引導您書寫一封完整的商務e-mail，教您如何善用電子郵件，讓e-mail成為您無往不利的溝通工具。

　　一般而言，一封e-mail包含以下幾個主題：

　　一、 Heading（信頭）

　　二、 Salutation（稱謂）

三、 Body of letter（正文）

四、 Terms of respect（敬辭）

五、 Signature（簽署）

六、 Postscript（附註說明）

　　在一封專業的商務e-mail郵件中，基本上都必須要具備上述的主題架構，每一種主題都有不同的表達重點，必須先瞭解其中的內容定位及書寫步驟。下一頁是一封商務e-mail的結構範例，讓您清楚地瞭解英文商務e-mail郵件的主題基礎結構。

商務英文 e-mail 結構範例

日 期	Date Sent: Tue, 5 May 2009
寄件者	From: Susan <Susan@yahoo.com>
主 旨	Subject: The new order
收件者	To: David <david@yahoo.com>
副 本	CC: John <john@yahoo.com>
附 件	Attachments: OP2009.doc
稱 謂	Dear David,

正 文

We've not received your order since last October. We need to find out what the trouble has been.

It's our policy to render the best service to our customers. You have always been considered as one of our regular customers, for you have given us remarkable patronage. We hope to have the pleasure of serving you again soon.

Your kind reply will be much appreciative.

敬 辭

Sincerely yours,

簽 署

Susan

附 註

P.S. We need you and we don't want to lose your business.

①

信頭

Heading

　利用e-mail作為商務書信的溝通管道時，首先，你必須要瞭解每一種e-mail工具所代表的意義，在撰寫商務書信時，才能更事半功倍地發揮其商務的功效。

　在一般的e-mail發信軟體中，都會有一制定的"Heading"(信頭)格式，雖然各家郵件軟體的格式不盡相同，但是所提供的功能是大同小異的，您可以依照個人的使用習慣或公司的安排，使用不同的郵件軟體做為溝通工具。

　以下這些信頭的內容代表的意義皆不同，您不得不了解其重要性。

一、　Date Sent（寄件日期）

二、　From（寄件者）

三、　Subject（主旨）

四、　To（收件者）

五、　CC（副本）

六、　Attachment（附件）

一、Date Sent（寄件日期）

"Date Sent"代表「發信件的日期」。已知目前市面上常見的電子郵件軟體（例如outlook、outlook express、yahoo、hotmail…等）幾乎都會在寄發郵件時，自動顯示發信當時的日期，有的電子郵件軟體甚至還會額外再增加發信當時的時間。

雖然不必親自註明發信的日期，但是你還是要瞭解在書寫英文商務信件時，日期的表達方式：

（一）日期的表達方式：

一般美式英文日期的書寫方式為："Tue, May 5, 2009"，表示是在「二〇〇九年五月五日星期二」所發的信函。

> ※常見的英文日期表達方式：
>
> 在英文書信中，常見的日期表達方式，有以下兩種方式：
>
> 1. "mm/dd/yy"，表示「月份/日期/年份」，例如：
>
>   05/16/2009
>
>   二〇〇九年五月十六日
>
> 2. "dd/mm/yy"，表示「日期/月份/年份」，例如：
>
>   16/05/2009
>
>   二〇〇九年五月十六日

因為東西方語言使用的慣例不同，又以「月份」和「日期」最容易為東方人所誤解，例如，若是書寫"04/05/2009"，那麼到底是「四月五日」還是「五月四日」呢？

建議在書寫商務e-mail時，若要特別提及日期時，「月份」最好不要用阿拉伯數字表示，而改以月份的英文全名(或縮寫)表達，以免月份和日期的表達產生混淆。

月份		
中文月份	英文月份	縮寫
一月	January	Jan.
二月	February	Feb.
三月	March	Mar.
四月	April	Apr.
五月	May	May
六月	June	Jun.
七月	July	Jul.
八月	August	Aug.
九月	September	Sep.
十月	October	Oct.
十一月	November	Nov.
十二月	December	Dec.

至於「日期順序」的表達方式，在正式英文商務書信中，日期的使用順序分別為：「星期」→「月份」→「日期」→「年份」，例如「二〇〇九年五月廿日星期四」，則可以有以下三種書寫方式：

(1) Thu, May 20, 2009

(2) Thu, May 20th, 2009

(3) May 20, 2009

上述第三種書寫方式，是不含星期說明的說明。

（二）星期的相關單字：

當你在說明日期時間時，若能夠同時提醒對方「星期」的日期，比較能夠幫助對方理解正確的時間點。若時覺得星期的說明文太長，也可以使用縮寫的說明。

星期		
中文星期	英文星期	縮寫
星期一	Monday	Mon.
星期二	Tuesday	Tue.
星期三	Wednesday	Wed.
星期四	Thursday	Thu.

星期五	Friday	Fri.
星期六	Saturday	Sat.
星期日	Sunday	Sun.
一星期		week
上班日、普通日		weekday
週末		weekend

　　此外，因為歐美人士對「星期」（week）的起始日是從星期天（Sunday）開始，而東方人普遍認知是從星期一（Monday）開始一週的計算點，所以當你提及「下週日」這個時間點時，最好搭配正確的日期說明，以免彼此對時間產生錯誤的認知。

※有關week的表示法	
上週	last week
上上週	the week before last
下週	next week
下下週	the week after next
每兩週	every two weeks
整整一週的時間	a week of days

(三)日期的表達方式：

　　日期是可以直接用數字表示，例如「五月廿五日」，就是用"25, May"表示。但在正式英文中，月份的日期應該是用「序號」表示，例如上述句子的「五月廿五日」，若用正式的英文書寫，則應該為"25th of May"，此外，要記住不可以用"May 25 "表示。

日期		
中文日期	英文日期	縮寫
一日	1st	first
二日	2nd	second
三日	3rd	third
四日	4th	fourth
五日	5th	fifth
六日	6th	sixth
七日	7th	seventh
八日	8th	eighth
九日	9th	ninth
十日	10th	tenth
十一日	11th	eleventh
十二日	12th	twelfth

十三日	13th	thirteenth
十四日	14th	fourteenth
十五日	15th	fifteenth
十六日	16th	sixteenth
十七日	17th	seventeenth
十八日	18th	eighteenth
十九日	19th	nineteenth
廿日	20th	twentieth
廿一日	21st	twenty-first
廿二日	22nd	twenty-second
廿三日	23rd	twenty-third
廿四日	24th	twenty-fourth
廿五日	25th	twenty-fifth
廿六日	26th	twenty-sixth
廿七日	27th	twenty-seventh
廿八日	28th	twenty-eighth
廿九日	29th	twenty-ninth
卅日	30th	thirtieth
卅一日	31st	thirty-first

※序號的使用原則

　　英文月份的日期的說明文並不是單純用數字表示，而是用數字的序號表示，例如「一月五日」，就表示是一月份的「第五天」所以是"5th of January"，因此英文日期的數字縮寫就是用序號表示。而上述日期的縮寫中，某些序號是比較不同的：

　　⇨1st = first（第一）
　　⇨2nd = second（第二）
　　⇨3rd = third（第三）
　　⇨11th = eleventh（第十一）
　　⇨12th = twelfth（第十二）
　　⇨13th = thirteenth（第十三）
　　⇨21st = twenty first（第廿一）

(四)年份的表達方式：

英文的日期可以直接用阿拉伯數字表示，例如「二〇〇九年」可以書寫為"2009"。

※「西元年份」的英文怎麼念？

　　1995 年的英文是"nineteen ninety-five"，字面解釋是「十九」加上「九十五」，那麼「2009年」怎麼說呢？可不是"twenty nine"，正確的說法應是"two thousand nine"，也就是「二千」加上「九」的說法，可不要說錯喔！

以下再總整理出所有日期的使用範例，您可以依實際需要改寫：

☑We'll place an order before September of 30, 2009.

☑We'll place an order before September 30th, 2009.

☑We'll place an order before Sep. 30, 2009.

☑We'll place an order before Monday, Sep. 30, 2009.

☑We'll place an order before the 30th of September.

二、From（寄件者）

"from"是「從～（何處）而來」的意思，代表是從某位「發信人」（sender）而來的信件，這是欄位的註明方式。

和「日期」一樣，在撰寫郵件時，軟體會自動加上你所註冊的帳號或顯示的名字。

若你是以公司的名義撰寫商務書信，建議應以讓對方在第一時間就瞭解你(或企業)的身份為「寄件者」的顯示設定，以免被收件者誤當成「垃圾郵件」(spam) 遭到刪除(delete)，而讓商務信件遭受石沉大海的命運。

因為"from"的內容是依個人各自的設定而有不同的名字，您可以選擇以下的顯示方式來表明你的身份：

(一) 表示「個人」的寄件者

範例 From: Susan<susan@yahoo.com>

來自 Susan，表示「個人名字」＋「郵件地址」，那麼收件人就可以知道寄信者使用哪一個帳號發信。

(二) 表示「身份全名」的寄件者

範例 From: Chris Jones

來自 Chris Jones，表示「名字＋姓氏」，而沒有顯示郵件地址」。

(三) 表示「公司代表」的寄件者

範例 From: Chris of BCQ

來自 BCQ 公司的 Chris，表示「個人名字」＋「公司名字」，清楚明瞭 Chris 所代表的公司。

(四) 表示「所屬企業部門」的寄件者

範例 From: BCQ Inc. Customer Service

來自 BCQ 企業的客戶服務部門，表示「公司的部門身份」，卻沒有註明個人的身份。

(五) 表示「公司身份」的寄件者

範例 From: the BCQ Product Upgrade Team

來自 BCQ 公司的產品升級小組，表示「公司的組織的身份」。

三、Subject（主旨）

　　"subject"表示「主旨」，代表這封商務書信的發信「主題」。

　　在分秒必爭的商業世界中，「時間」是非常珍貴的，而一封有效率的商務書信，其「主旨」（subject）欄位所傳達的內容，更是能夠幫助「收信人」（receiver）對信件的篩選產生相乘的作用，自然能夠讓你的e-mail能夠在一堆「垃圾郵件」（spam）中脫穎而出。

　　此外，也可以用"Sub"或"Re"表示主旨的說明文開頭，前者是"Subject"的縮寫，後者是"Reply"的縮寫，適用在「回覆」對方前一封信件的意思。

　　另外要特別注意的是，若屬於正式書信，則應避免使用像是"Hi"、"FYI"或是"One more thing..."的字眼，上述這種語焉不詳或是過於隨性的主旨用語都是禁忌。

（一）單字主旨

　　以下是一些在商務書信中常見的「主旨」的字詞用法，提供您簡化書信內容的解釋：

　　　1.Quotation　　　　報價單、估價單

　　　2.Offer　　　　　　出價、報價

3.Bargain 議價

4.Budget 預算

5.Product 商品

6.Sample 樣品

7.Catalogue 型錄

8.Purchase order 訂單

9.Contract 合約

10.Agreements 協議

11.Shipment 裝船、船運

12.Damage 損壞

13.Shortage 數量短缺

14.Complaint 抱怨

15.Reminder 提醒

16.Agency 代理權、代理商

17.Payment 付款

18.Inquiry 詢問

19.Request 需求

20.Information 資訊

21.Follow-up 後續追蹤

22.Detail 細節

23.Proposal 企畫書

若是一個單字詞無法完整說明您的商務需求，則以下兩個字詞的主旨也可以提供您使用：

1.Business Expansion	商務拓展
2.Trade Proposal	交易提案
3.Agency Application	徵求代理
4.Business Outline	商務概況
5.Business Introduction	商務介紹
6.Statements of Accounts	財務報表
7.Proposal Accepted	接受提案
8.Proposal Rejected	拒絕提案
9.Sales Contract	買賣合約
10.Inguiry for Textile Goods	紡織品尋價
11.Price List	價目表
12.Placing an Order	下訂單
13.Order Confirmation	訂購確認書
14.Request for L/C	要求開立信用狀
15.Description of goods	商品說明
16.Making an appointment	安排會議
17.Looking for Customers	尋找客戶
18.Extending Business	拓展業務
19.Selling Offer	賣方報價
20.Buying Offer	買方報價

（二）簡短句型的主旨

此外，也可以利用較長的「主旨」說明，以提醒收信者注意您的商務書信，但是要注意句子的長度，仍以簡易句子為原則，以免過於冗長，例如以下都是不錯的主旨說明：

範例 Lunch rescheduled to Friday @ 1pm
午餐改至星期五的下午一點鐘

範例 Reminder: Monday is "Thanksgiving Day"-no classes
提醒，星期一是感恩節，不用上課

範例 HELP: Can you revise my proposals?
請幫忙：可以幫我修正我的提案嗎？

※備註：主旨若是都用大寫表示，目的是要提醒收信人「特別注意」你的這一封來信的重要性。

範例 Thanks for the new solution - works great!
感謝新的解決方案，的確有效！

（三）「主旨＋說明」的表達方式

商務書信的主旨內容也可以是「主旨」＋「日期」或是「主旨」＋「說明」的方式表示。

範例 Sub: Your letter of Jan. 30, 2009
主旨：您二○○九年一月卅日的來信

範例 Sub: Your Order No. 415
主旨：您編號415號的訂單

範例 Sub: Inquiry from BCQ
主旨：來自BCQ的詢問

（四）「回覆＋說明」的表達方式

可以利用前一封發信者的主旨，直接回覆給對方，那麼你的e-mail的主旨中，就會先顯示出對方的主旨，並在最前方出現"Re:"的顯示，表示「回覆信件」的意思。

範例 Re: Shortage of your goods
回覆：您短缺的商品

範例 Re: Sorry for being late
回覆：為遲到而抱歉

範例 Re: Placing an order
回覆：下訂單

四、Receiver（收件人）

"receiver"代表這封商務書信的寄發(send)對象。商務書信中，必須有一特定的寄件對象(也就是「收件人」)，在您選擇對方的郵件地址之後，對方的名字(或郵件地址)就會出

現在此欄位中。

　　一般而言，每一種寄信軟體都有「通訊錄」(address book)，在設定「收件人姓名」的內容時，就可以在其名字之後，直接註明對方的公司、職銜，而郵件地址(e-mail address)就鍵入至相關的欄位中，當然也可以僅用其名字顯示即可，但是前提必須是此郵件軟體有「顯示名稱」的選項。

範例 To: "Mr. White"

給「Mr. White」，直接表示收信者姓名。

範例 To: "Chris"chris@yahoo.com

給「帳號 chris@yahoo.com 的 Chris」，表示「個人名字」＋「郵件地址」的收信對象。

範例 To: "Chris"of BCQ

給「BCQ公司的Chris」，表示「個人名字」＋「公司名字」的收信對象。

範例 To: "Chris Jones"

給「Chris Jones」，表示「名字＋姓氏」。

範例 To: "Manager of Marketing Department"

給「行銷部門經理」，表示「職銜」。

五、CC（副本）

當需要將此信寄給第三者時，就可以在"CC"(副本)的欄位中鍵入第三者的郵件地址，而主要的收信人也會知道你同時有給第三者這一份信件的副本。

收件者	To: David <david@yahoo.com>
副　本	CC: John <john@yahoo.com>
稱　謂	Dear David,

"CC" 的 全 文 是"Carbon Copy"。"Carbon" 的 原 意 為「碳」，因為以前的複寫紙都含有碳，所以「複寫的副本」的英文就叫做"Carbon Copy"。

此外，若是需要將此封e-mail寄給第三者，但又不想讓主要收信人知道有第三者也收到這封信，您就可以在"BCC"的欄位中鍵入第三者的郵件地址。"BCC"的全文是"Blind Carbon Copy"。"Blind"是表示「視而不見的」，藉此表示收件者看不見副本收件人(「看不見」就表示「不知道」)，這封e-mail也就成為「密件副本」，表示只有寄件者知道此「密件副本」寄給了哪些人的意思。

六、Attachment（附件）

除了選取「附加檔案」功能加入檔案之外，你也可以在內文當中說明您有隨信附加檔案給對方，以提醒對方開啓e-mail信件中所附加的檔案，也避免對方將你的附件檔案誤認爲夾帶病毒的郵件而刪除。

> **範例** Attached please find a submission for your consideration.
>
> 附件為您請你所需要的協議書。

> **範例** I have attached the new Word document.
>
> 我已經附寄了新的Word檔案。

> **範例** The new Word document is attached.
>
> 新的Word檔案已經如附件。

> **範例** Please take a look at the attached Word document.
>
> 請詳見附件新的Word檔案。

此外，因爲電腦病毒猖獗，所以也要避免將副檔名爲exe的執行檔當成附件寄給對方，以免引起對方的不安而導致不敢開啓您的e-mail或附件檔案。

②

稱謂

Salutation

"Salutation"「稱謂」是屬於e-mail信件中,在書寫正文前的第一部份,也就是中文書信「敬啓者」的意思。

"Salutation"「稱謂」代表對收件人的尊稱,就如同與人見面要打招呼一樣,不可不注意在此時應注重的禮儀,特別是商務書信中,您所使用的所有稱呼用詞都攸關著是否能成功地利用「e-mail商務書信」作爲溝通管道。

由於英文是東西向橫寫文字,「稱謂」的書寫位置便在信件本文的第一行,必須靠齊內文的最左側。

不管是正式的或非正式的商務書信,「稱謂」的使用都應以尊敬對方、與對方保持良好關係爲前提,以下是一些商務書信中常用的稱謂方式:

(一) Dear Sir「敬啓者」:

用在「只知其人不知其名」的狀況下,例如想要向某公司的「客戶服務部門」投訴時,因爲你不會知道是哪一個特定人員閱讀此封e-mail郵件,就可以用此稱謂尊稱所有可能

的收信者。

(二) My dear Sir「敬啟者」：

　　也是用在「只知其人不知其名」的狀況下，使用情況和"Dear Sir"相同，但是多了"My dear"的稱呼，便多了一份對收信者的尊敬、謙虛的態度。

(三) Dear Sirs「敬啟者」：

　　用在只知爲一群人中的「某一人」，使用情況也類似上述情形，但是你確知那可能會是「很多人」都會收到訊息的情況下時使用，"Sirs"表示複數的一群人的意思。

> ※sir的性別？
>
> 　　一般而言，"sir"多半適用在男性對象，但是若是不知收信者的性別時，不論男女，有時也會統一使用"sir"的尊稱，此時便不必在意對方的性別，但前提必須注意，對方是你所不認識的人才能使用。

(四) Dear Madam「親愛的小姐」：

　　"Madam"表示「夫人」、「太太」、「小姐」的尊稱，適用對象不分已婚或未婚，也不知該名女士的稱呼，但是表示你已知收信者是爲「女性」，但不知道對方的身份或姓名，也無從得知是哪一位特定女性會閱讀郵件。"Dear Madam"是正式書信中，對女性的最常用的尊稱，"Madam"後面不加任

何的姓氏或名字。

(五) Dear Sir [Madam]「敬啟者」：

　　在書寫e-mail時，男女的性別都註明時，就有點類似中文所說的「先生或小姐」的意思，也是適用在不知收信者性別的情況下使用，因此不論收信者是哪一個性別，都不會顯得突兀、不禮貌。

(六) Dear Mr. /Mrs. Jones「親愛的瓊斯先生/瓊斯太太」：

　　"Mr."是"Mister"的簡稱，而"Mrs."則為"Mistress"的簡稱。

　　"Mr./Mrs. ＋姓氏（last name）"用在僅知道對方的姓氏，適用於所有正式場合或非正式場合對已知身份的男性的尊稱。

　　但是若對女性使用"Mrs."，則是確認對方為「已婚身份」，並是冠上「丈夫姓氏」時使用，所以"Mrs.＋丈夫姓氏"是表示「某某夫人」的意思，例如稱呼美國總統歐巴馬(Barack Obama)的夫人蜜雪兒時，就應該使用"Mrs. Obama"(歐巴馬夫人)。

　　若是您不知對方（女性）是否已婚，或是是否有冠上丈夫的姓氏，建議您還是用"Dear Madam"比較不會造成雙方的誤解。

※ Mr./Mrs./Ms.的用法

　　Mr./Mrs./Ms.均是尊稱的用法，類似中文「先生」、「太太」、「小姐」的意思，適用的對象不同。

(O) Mr./Mrs./Ms.＋姓氏

範例		
Mr. Jones	瓊斯先生
Mrs. Jones	瓊斯女士/太太
Ms. Jones	瓊斯小姐

(O) Mr./Mrs./Ms.＋名字＋姓氏

範例		
Mr. John Jones	約翰・瓊斯先生
Mrs. Susan Jones	蘇珊・瓊斯女士/太太
Ms. Judy Jones	茱蒂・瓊斯小姐

(X) Mr./Mrs./Ms.＋名字

　　不可以單獨將名字（first name）和姓氏尊稱合用，例如"Mr. John"就是錯誤的用法。

(七) Dear Mr. and Mrs. Jones「親愛的瓊斯先生及瓊斯太太」：

　　代表收信者是一對已婚的夫妻，並且信件內容與這對夫妻都有關係時適用，像是邀請夫妻一同參加聚會、共同通知夫妻雙方某事時，就可以使用"Dear Mr. and Mrs. Jones"。要注意的是，這裡的姓氏要冠上男性的姓氏（last name），而非女性的姓氏。

※Mr. and Mrs. 夫婦的用法

(O) Mr. and Mrs.＋姓氏

範例　Mr. and Mrs. Jones　瓊斯夫婦

(O) Mr. and Mrs.＋先生名字＋姓氏

範例　Mr. and Mrs. John Jones　約翰‧瓊斯夫婦

(X) Mr. and Mrs.＋太太名字＋姓氏

不可以在使用"Mr. and Mrs."時，冠上女士的姓氏(last name)或名字（first name），例如"Mr. and Mrs. Mary Jones"（瑪莉‧瓊斯夫婦）就是錯誤的用法，因為"Mr. and Mrs."是指夫妻雙方的尊稱，而非單指女士的尊稱。

(八) Dear Ms. Jones「親愛的瓊斯小姐」：

　　若是對方為未婚的女性，並知道其姓氏時，就可以使用"Ms.＋姓氏"稱呼對方，表示對未婚女士的稱呼，"Ms."是"Miss"的縮寫。此外，有時明知對方為已婚身份，也可以用"Miss＋（女性）姓氏"的方式稱呼對方，但似乎會顯得刻意逢迎的態度。

※Mr. and Mrs.的動詞使用規則

　　要特別注意的是"Mr. and Mrs."的動詞使用規則，"Mr. and Mrs.＋姓氏"，代表兩人的身份，所以是使用動詞的複數規則。

> **範例** Mr. and Mrs. Jones were married on September 25, 1965.
>
> 瓊斯夫婦是在一九六五年九月廿五日結婚。

(九) Dear Ms. Jones「親愛的瓊斯女士」：

　　當你已知對方身份，也知她的姓氏，但表示無法確定對方是否已婚或未婚的女性身份，或對方為不願提及婚姻狀況的女性時，也可以用"Ms.＋姓氏"來稱呼對方，簡而言之，就是已婚、未婚女士皆適用。

(十) Dear Doctor Jones「親愛的瓊斯醫師」：

　　當你已知對方的職銜時，就可以用"職銜＋姓氏"稱呼對方。無關乎男女性別或是否已婚的身份，"職銜＋姓氏"的稱謂方式都適用在商務書信中，而職銜第一個字母通常都使用大寫格式。(職稱詳見P59說明)

(十一) Dear Customer「親愛的客戶」：

　　當只知對方身份但沒有特定哪一位對象時，就可以使用"Dear＋身份"的稱謂，通常是對對方表示無上的敬意時使用，像是給顧客（Customer）、股東（Stockholder）、會員（Member）、出席者（Attendant）、員工（Employee）的e-mail都可以採用"Dear＋身份"的方式，此種信函也有類似

「通知函」的功能；同上述職銜，第一個字母也是大寫格式。

(十二) Gentlemen「敬啟者」：

　　使用情況和"Dear Sirs"類似，適用在你的寫信目標為只知為「某一群人」時，在商務書信中的使用非常普遍。"Gentlemen"是「紳士」（Gentleman）的複數名詞。

(十三) To whom it may concern「貴寶號鈞鑒」：

　　當你寫信給某一公司或單位時，在不知道收件者是某位特定人、特定部門時，就可以使用"To whom it may concern"，字面意思是表示「給予此事相關者」，即中文「貴寶號鈞鑒」或「敬啟者」的意思。

(十四) Dear Chris「親愛的克里斯」：

　　直接稱呼對方"Dear＋名字"是使用在非常熟悉對方的情況下，其熟悉的程度是雙方根本可以稱得上是「朋友」的關係或熟識多年的客戶等。

(十五) Dearest Chris「給最親愛的克里斯」：

　　適用在非常親密的朋友、親人、伴侶之間的用語，很少人會使用在正式的e-mail商務書信中，否則便有過於拉攏關

係、拍馬屁的嫌疑。

(十六) Hi Chris「嗨，克里斯」：

　　只是簡單地用"Hi"打招呼的方式，是與對方有一定的交情的情況下，是一種非常隨性的稱呼方式，在某些正式的 e-mail 書信中並不建議使用。

　　商務 e-mail 並不是都是嚴肅的，若您與對方已經有一定的熟悉程度，偶爾將嚴肅而沒有感情的稱謂改為用一種輕鬆而不失禮的方式表現也是被允許的。

3

「正文」
Body of letter

「正文」(Body of letter)是書寫這一封「e-mail商務書信」的宗旨精神。"body"是「身體」之意，「信的身體」也就是「正文」之意。

一封好的商務書信，必須能夠發揮其應有的商業開發、溝通、解決問題的功能，這一部份也是收件人決定是否要保留或注意這封e-mail的「關鍵時刻」。

商務書信最忌諱的就是無關緊要的廢話，在撰寫商務e-mail書信時，你必須提出幾個問題先自我回答：

「內容的重點為何？」

「希望對方如何重視這個問題？」

「解決的方法為何？」

並不是用最艱澀、深奧的英文就表示自己的英文能力一

流，因為如此一來，不但閱讀不易，反而容易讓客戶產生你「很難相處」、「容易刁難」、「炫耀英文能力」的錯覺，因此，一封真正有效率的商務e-mail是能夠用簡單的英文，就達到雙向溝通的功能。

此外，也切忌將商務書信當成閒聊的工具，把握"key word"「關鍵用詞」的使用，才是書寫一封成功商務e-mail的訣竅。

一般而言，不管是中文或英文的商務書信，在信件的「正文」(Body of letter) 中，都包括了三大「段落」(paragraph) 的內容：

一、Opening「開場白」

二、Middle「中間主文」

三、Complimentary Close「結束語」

根據不同的商業訴求，每一段落都必須扮演好各自的角色，才能有效地發揮商務e-mail的溝通功能。以下就這三大段落做一個詳細的使用說明介紹：

一、開場白

在e-mail開場白一開始就簡單地打招呼，可以拉近彼此的距離，幫助您達成商務e-mail的首要任務。

第一段的"Opening"「開場白」內容除了具備「簡單地向
對方打招呼」的功能之外，也可以將你在之後的中間主文
(Middle) 所要詳述的內容先做簡單說明，通常以簡單一兩句
說明即可帶出重點。

一般而言，這一段落有以下四種主題句型可以發揮使用：

（一）簡單打招呼

利用簡單的打招呼的方式作為開場白，以免信件一開始
就是嚴肅的商務話題。

> **句型** It's been a long time since ... 自從～後已經
> 很久了
>
> **範例** It's been a long time since we met in Hong
> Kong.
> 從上一次我們在香港見面後，已經好長一段時
> 間了。

> **句型** How is ...? ～好嗎？
> **範例** How is your business?
> 生意好嗎？

（二）已知對方的現況：

表示已經約略地知道對方的現狀後，告知對方自己已經
獲知某事，以表達關心之意。

句型　We heard from ... 我方從～(某人)得知

範例　We heard that promotion from Mr. Smith.
我們從史密斯先生那兒得知此升遷一事。

句型　We heard of ... 我方得知～一事

範例　We heard of your final decision.
我方得知您的最後決定。

句型　We're terribly sorry for ... 關於～一事，我方感到很抱歉

範例　We're terribly sorry for the shortage.
關於短缺一事，我方感到很抱歉。

句型　We regret to know ... 很遺憾得知～一事

範例　We regret to know the decision of your company.
很遺憾得知貴公司的決策。

（三）已收到對方的來信：

主要是告知對方，你已經收到他前一封的來信了，至於你會如何處理，則尚未特別說明。

句型　Thank you very much for your letter dated ...
感謝您於～（日期）的來信

範例　Thank you very much for your letter dated 25th of January.
感謝您於一月廿五日的來信。

> **句型** We're pleased to receive your letter about...
> 很高興收到您關於～的來信
>
> **範例** We're pleased to receive your letter about the quotation.
> 很高興收到您關於報價的來信。

> **句型** As requested in your letter, ...　根據您來信要求～
> **範例** As requested in your letter, we'll consider your proposal.
> 根據您的來信要求，我方會考慮您的提案。

（四）回覆對方的詢問：

　　表示對方曾經來信給你，而你現在就上次對方的來信主題回信給對方。

> **句型** In response to your letter, ...　回覆您的來信～
> **範例** In response to your letter, we suggest to you that we should meet.
> 回覆您的來信，我方建議我們雙方應該見面。

> **句型** As regards to ...　關於～一事
> **範例** As regards to your suggestion, we'll reschedule this appointment.
> 關於您的建議，我方會將這個約會改期。

> 句型 In connection with ...　關於～一事
> 範例 In connection with that offer, Mr. Smith will call you back.
> 這是有關於報價一事，史密斯先生會回您電話。

二、中間主文

　　第二段的"Middle"「中間主文」內容就是這一封商務e-mail書信的「主題」(subject)，也是成就您的商業行銷策略是否能成功策馬入林的重要關鍵，在開場白簡單打過招呼後，您就應該盡快進入主題，千萬不要再提一些與主題無關緊要的旁枝末節的事件。

　　以下提供幾點書寫商務書信的重點建議，是您在撰寫此段書信的重要參考方式：

　　（一）重點：簡潔有力的文字敘述，避免過於冗長及艱澀的說明。

　　（二）話題：提供對方感興趣的話題。

　　（三）問題：提出問題以吸引對方注意。

　　（四）保證：提供有利的證據、保證或輔佐資料。

範例 While we're very interested in your qualifications, we're afraid that you don't quite meet our needs.

儘管我方對貴公司的條件感到很有興趣，不過恐怕貴公司不符合我方的需求。

範例 Our annual requirements for household products are considerable, and we may be able to place orders with you if your prices are competitive and your delivers prompt.

我們每年的家庭用品需求量極大，如果貴公司價格公道、交貨迅速，我方將會下訂單。

範例 We're very sorry to inform you that your last delivery is not up to your usual standard.

很遺憾通知，貴公司前一次的商品品質不如從前。

範例 Now your settlement is 2 months overdue and we look forward to receiving your remittance within a week.

如今貴公司的付款期已超過二個月了，我方希望能於一週內收到匯款。

　　以上的撰寫方式都是為了成功地達成商務email在行銷時的助力，但是最終的目標，仍是以能成功引起對方的「注

意」(Attention)，並發揮「實際行動」(Action) 的功效。

三、客套結束語

　　一封成功的商務書信要能前後連貫，「主文」是從收件人的立場為書寫的觀點，而第三段的「客套結束語」則站在寫信人的立場結束，因此必須簡潔有力，通常只要一句話說明即可，具有「靜待佳音」、「靜待您的回覆」或「不甚感激」的意思，旨在傳達您的耐心等待及誠意。

　　以下是一些常見的結束語的句型及範例，可以幫助您更快得到對方的回覆：

（一）期待語句：

　　期待對方能針對你的e-mail而能有所回應，而你也會等待對方的來信說明。

> **句型** be looking forward to ... 期待～
>
> **範例** We're looking forward to receiving your reply.
> 我方期望得到您的回覆。
>
> **範例** I'm looking forward to your comments.
> 我將靜待您的建議。

（二）感謝語句：

主要是感謝對方，不管是因為對方的來信或提供的幫助等，所謂「禮多人不怪」就是這個意思，讓對方感受你所釋出的善意。

句型 Thank you for ... 感謝～

範例 Thank you for your cooperation with us in this matter.
感謝您對於此一事件給予我方的協助。

範例 Thank you again for your attention.
感謝您對於此事的注意。

（三）要求語句：

有事有求於對方，希望對方能夠配合告知你相關訊息。

句型 Please inform/tell us of ... 請告訴/通知我方～
（某事）

範例 Please inform us of your decision soon.
請盡快通知我方您的決定。

範例 Please tell us whether you may accept it.
請告知我方您是否會接受。

（四）詢問語句：

適用於詢問的主題，有一點類似要求語句，能和任何的動詞

做搭配使用，表達謙虛的詢問主題。

> **句型** Will you please ...?　　能否請您〜？
>
> **範例** Will you please reply us before this Wednesday?
> 能否請您盡快在這個星期三前回覆？
>
> **範例** Will you please send us a copy of your latest catalogue?
> 能否請您寄給我方一份貴公司的最新目錄？

（五）堅信語句：

　　針對某一事件，表達你所保持的堅定認知，期望對方能夠被你所說服。

> **句型** We trust/believe/await ...　我方相信/堅信/等待〜
>
> **範例** We trust you will now attend to this matter without further delay.
> 我方相信貴公司會毫不延誤地關切此事件。
>
> **範例** We await your satisfactory to our quotation [service/product].
> 我們等待貴公司對我們的報價[服務/商品]感到滿意！

（六）謙虛語句：

　　表示對對方所提的問題完全開放接受，並希望對方不必拘泥世俗的客套推託，儘管放心地提出問題。

| 句型 | You're welcome to ...　歡迎做某事～ |

| 範例 | You're welcome to ask BCQ Company for any help. |
| | 歡迎您隨時要求BCQ公司提供任何的協助。 |

| 句型 | You're the most welcome to ...　歡迎你做～ |
| | (某事) |

| 範例 | You're always the most welcome to contact us. |
| | 歡迎您隨時與我方聯絡。 |

（七）保持聯絡語句：

　　希望雙方能夠持續保持聯絡，讓對方能夠放心地將你當成關係良好的事業伙伴。

句型	Please don't hesitate to ...　請不要遲疑去做～
範例	Please don't hesitate to contact us.
	不要遲疑，請馬上與我們聯絡。

句型	Feel free to ...　不要客氣去做某事～
範例	If you have any questions, feel free to call me.
	如果您有任何問題，不要客氣打電話給我。

4

敬辭
Terms of respect

「敬辭」視為一個句子，所以必須使用大寫的格式。

有點類似中文「謹啟」、「敬啟」的意思，通常有以下幾種表現方式，由上至下依序為「最尊敬」到「普通程度」的敬辭使用：

Very sincerely yours,

Sincerely yours,

Respectfully,

Faithfully yours,

Yours very truly,

Very respectfully yours,

Yours,

Best regards,

Best wishes,

Warmest regards,

Regards,

　以上都是屬於較正式的用法，若是你和對方已經是相當熟識的伙伴，則可以採用以下較爲隨性的用法，免得顯得太過於生疏或正式：

　Take care, 「保重」

　All the best, 「祝好」

　Be good, 「祝好」

　Cheers,「祝愉快」

5

簽署

Signature

從前只有紙張可以書寫的商務書信時，「簽署」這一部份就包括「寫信人」及「發信人」的簽名，如今 e-mail 盛行，「代爲書寫文書」的情形根本不存在了，因此在此是以「發信人」的立場來解釋「簽署」的作用，主要有「發信人姓名」及「職銜」、「公司」三大部分。書寫的位置是以齊信件的左邊邊界爲主。

雖然在信件的信頭可以看見發信人的身份或名字，但還是建議您能夠正式地告訴對方您的名字、職銜或公司名稱。以下幾種方式您可以自行選擇使用：

一、只書寫發信者的[名字]

在信件結尾處只寫上你的名字的作法，表示您和收信者有一定程度的熟識度，則可以在 e-mail 最後署名處，註明自己的名字即可，對方一看就會知道你的身份。

> **範 例**
>
> If you have any questions, feel free to call me.
>
> Best regards,
>
> Chris

二、書寫[名字＋姓氏]

　　若註明發信者的[名字＋姓氏]，則表示您和對方不是那麼熟識，也擔心對方不確定您到底是哪一位（因為有可能同名的機會很大），所以完整的書寫自己的[名字＋姓氏]，以方便對方辨識您的身份。

> **範 例**
>
> Please don't hesitate to contact us.
>
> Yours very truly,
>
> Chris Jones

三、書寫[姓名＋職稱]

　　若是屬於正式的商務e-mail信件，那麼建議您在名字之後加上自己的職稱(職稱的英文名稱，請詳見P59)，以表示自己發這封信是以公司任職的身份為代表。要記住，和中文信件不同，英文信件中，職稱的位置是在姓名之後。

範 例

If you have any questions, feel free to call me.

Yours very truly,

Chris Jones

Assistant Manager

四、書寫[姓名＋職稱＋公司名稱]

除了身份職稱之外，也可以將自己所代表的企業或組織一併註明，如此一來，將有助於對方在第一時間就知道您所代表的單位。

範 例

If you have any questions, feel free to call me.

Yours very truly,

Chris Jones

Assistant Manager

BCQ Co. Ltd.

五、書寫[姓名＋職稱＋公司名＋聯絡方式]

除了身份之外，也可以將自己的聯絡方式（電話、傳真或e-mail）註明在署名處的最後，萬一收信者需要立即和你用電話聯絡，就可以馬上撥電話，這個貼心的動作，可以為收信者節省不少的時間。此外，若是屬於跨國之間的商務e-mail往來，在註明聯絡電話時，要將您自己本國的聯絡電話的國際碼一併註明，如此一來才是完整的聯絡方式。

範 例

If you have any questions, feel free to call me.

Yours very truly,

Chris Jones

Assistant Manager

BCQ Co. Ltd.

Tel:002-886-02-86473663

Fax:002-886-02-86473660

chris@foreverbooks.com

※職場職務名稱

行政管理類

董事長	president
總經理	general manager
管理者	administrator
主管	supervisor
秘書	secretary
企畫人員	project staff
人事經理	personnel manager
人力資源部經理	human resources manager
培訓師	trainer
生產經理	production manager
公共關係經理	public relations manager
高級顧問	senior consultant
品質控制員	QC inspector

工程技術類

程式設計師	programmer
電腦操作員	computer operator
系統分析員	systems analyst
系統操作員	systems operator
技術員	technician

建築工程師	architectural engineer
總工程師	chief engineer
電腦工程師	computer engineer
顧問工程師	consulting engineer
工程師	engineer
軟體工程師	software engineer

商業會計類

會計員	accounting
查賬員	auditor
簿記員	bookkeeper
出口人員	export clerk
外銷員	export sales staff
財務分析員	financial analyst
市場分析員	market analyst
市場研究員	market researcher
銷售代表	marketing representative
商品業務員	merchandiser
採購員	purchaser
業務經理	business manager
外銷部經理	export sales manager
業務助理	business assistant

6

附註說明
Postscript

"postscript"是「信末的附筆」的意思，此即為大家所熟知的"P.S."，針對當你完成書信後，補足其遺漏待說明的部分，或是當您必須中斷此文句的本文說明時使用，通常是在非正式信件中較常使用。

"P.S."的全文是"Postscript"，但是現在有越來越多的情況是利用"P.S."作為加強印象的功能。

> **範例** P.S. Did I mention she's coming to visit?
> 附註說明，我有提過她下星期會來拜訪的事嗎？

> **範例** P.S. Don't forget to visit www.foreverbooks.com
> 不要忘記去瀏覽www.foreverbooks.com網站。

此外，若是已經註明了"P.S."的說明文後，又要再補充說明，則可以用"P.P.S."再次說明。"P.P.S."的全文是"Post-Postscript"。

若要再補充第三個"P.S."呢？則可用"P.P.P.S."表示，並以此類推。

7

隨信附加檔案
Attached file

　　因為電腦「病毒」(Virus)常常隨e-mail所夾帶的檔案進入收件人信箱中，若是隨便開啓來路不明的附加檔案，電腦便有中毒的可能，因此在使用e-mail商務書信時，並不建議您附加(attach)任何的檔案。

　　若是不得不附加檔案給收信者時，您也有義務告知對方所附加的檔案為何種檔案，讓收信者可以放心開啓所附加的附件檔案。以下是幾種附件的書寫句型，以告知對方您隨信所附的檔案說明。

一、隨信附上

> **句型** Attached, please find ...　詳情請見附件～
>
> **範例** Attached, please find the new Annual Report file.
>
> 隨信附上新的「年度報告」檔案。

> **句型** Attached, you'll find ... 隨函可以找到～
>
> **範例** Attached, you'll find the new Annual Report file.
> 隨函可以找到新的「年度報告」檔案。

　　以上的說法都是屬於比較正式的說法，若是不想要這麼嚴肅的用法，也可以試試以下幾種方法，比較口語、簡單，適用在和對方已經有點熟識時使用：

> **範例** I have attached the new Annual Report file.
> 我已經附上新的「年度報告」檔案。
>
> **範例** The new Annual Report file is attached.
> 新的「年度報告」檔案已經附上。
>
> **範例** Please take a look at the attached Annual Report file.
> 請詳閱附件新的「年度報告」檔案。
>
> **範例** Replace the old Annual Report file with the new one which is attached.
> 替換舊的「年度報告」檔案，附上新的「年度報告」檔案。
>
> **範例** You'll find the new Annual Report file attached to this e-mail.
> 隨此封電子郵件，附上新的「年度報告」檔案。

二、以郵件寄出

　　若是需要在發e-mail之外另外將相關資料透過實體郵寄的方式寄給對方，也需要在e-mail中說明清楚。

句型 We'll send you ... 我們會把～寄給您

範例 We'll send you the contract separately.
我們會另外再把合約寄給您。

句型 We're pleased to send you ... 非常榮幸地寄～給你

範例 We're pleased to send you the 2008 Summary Financial Report.
非常榮幸寄給你二○○八年財務總結報告。

句型 Would you please send us ...? 能將～寄給我們嗎？

範例 Further to your request, would you please send us the new form first?
在此應您的要求，能否寄給我方新的表格？

句型 We'd appreciate it if you would send us ...
如果能將～寄給我的話，我們將感激不盡

範例 We'd appreciate it if you would send us the samples.
如果能將樣品寄給我方，感激不盡。

而一般人也都被教育成避免打開「附加檔案」(Attach-ment File)，所以建議您可以用「超連結」(hyperlink) 的方式指引收信者「點選」(click) 到相關「網頁」(website)。

> 範例　Please visit our sponsor: www.foreverbooks.com
> 請參觀我們的贊助商網站：www.foreverbooks.com
>
> 範例　Click here for more information about our services.
> 點選這裡，以得到更多我方的服務訊息。
>
> ※備註：大部分的郵件軟體都會在你加入郵件地址或網址時，自動產生連結至郵件軟體的書信畫面或網站連結，且文字會產生藍色的變化，以供連結之用。

※「點選網頁」要如何表示？

　　"click"是「卡嗒聲」，聽起來很像「用滑鼠點選」的聲音，所以「點選網頁」的英文就叫做"click website"。

此外，提供「免付費電話」(Toll-free) 的訊息也是一個方式，但是根據網路行為者的調查研究統計，一般人寧願點選網頁查詢資料，而懶得拿起電話撥打給客服部門(CSR)。

Part 2 簡潔的書寫訣竅

E-mail是透過電腦的網際網路，傳輸訊息（messages）或檔案（files）的管道。當電腦網路連接至網際網路時，員工就可以透過e-mail立即和一個街道外的人、甚至是世界另一邊的人溝通。

E-mail是一種非常方便的溝通工具，透過傳送、接受備忘錄、訊息的方式，使雙方能夠達到有效溝通的目的。也幸好有網路瀏覽器能夠夾帶附件的功能，我們才得以將冗長的文件，透過文字編輯的軟體（像是word或text的文件），傳送給對方，節省了大量的時間及郵資費用。

簡單、方便、有效的溝通工具

有時候，和正式的書信寫作比較起來，書寫一封e-mail似乎是一件非常簡單的事，但是商務e-mail也的確需要注意一些書信寫作的技巧。

E-mail已經成為大部分企業組織中，非常重要的溝通工具。原因為何？因為e-mail快速、有效率、簡單且費用精簡

的特色。

可惜的是，就是因爲這種快速、簡便的商務e-mail溝通模式，使得e-mail的應用對個人及企業組織產生了一些問題。

首先，像是垃圾郵件（spam）對系統所造成的資源、容量的負擔，以及員工花了太多心思在處理私人郵件上的現象…等等。

第二，過於氾濫的e-mail往返，也容易產生書寫e-mail時，所產生漫不經心的馬虎格式。人們通常只是將他們心中所思考的問題寫出來，便按下傳送按鈕（Send button），完全沒有對所書寫的e-mail做全盤思考，甚至是缺乏深思熟慮的思考態度。別忘了，若是對方有收到這一封e-mail並打開閱讀時，他可是會有極大的機率是一字一字閱讀著這一封不夠得體的商務e-mail，除非您期望對方忽略您這一封有瑕疵的商務e-mail。

第三，夾帶情緒性及思考不周的e-mail，往往在書寫者有時間冷靜下來時，就已經書寫完成並寄送出去了，這也是「快速溝通」所造成的危機。

第四，很多時候，不夠嚴謹的寄送過程，也容易將email錯發給非預期的收信對象（recipients），有時也因此產生負面的效果。

最後，縱使您已經將e-mail刪除（delete），也可以透過某些非法的手段，重新復原這些信件，也許這些重見光明信件的威脅程度，是會讓人丟工作的。

其實以上這些e-mail所產生的困擾，都是可以透過一套完整的書寫機制，避免大部分的問題產生。

書寫e-mail的訣竅
想要有效利用e-mail作為溝通的工具，就必須要特別用心在書寫e-mail時，不得不注意的一些重要秘訣。

一、主題明確
何謂「主題明確」？很簡單，就是讓對方收到e-mail後，一看到你的主旨列或是e-mail的開場白時，就可以一目了然，使對方能夠立即瞭解你所發的這一封e-mail所要傳達的重點。

(一) 開場白重點：期待建立商務關係
當你需要拓展商務關係時，以下的範例可以簡單地介紹公司，並進一步期望和對方建立合作關係。

範 例

> We're one of the importers of wooden furnitures in Taiwan and shall be pleased to establish business relationships with you.
>
> 敝公司為台灣的實木傢具進口商之一，很高興和貴公司建立商務關係。

(二) 開場白重點：要求對方提供報價資料

當已經準備建立合作關係時，免不了需要請對方提供一些相關資料。

範 例

> We're very interested in your cameras and digital cameras. Will you please send us a copy of your catalogue, with details of prices and items of payment?
>
> 我方對貴公司的相機及數位相機非常有興趣。貴公司是否能寄給我方一份貴公司的型錄，並附上價格、付款明細？

(三) 開場白重點：提供報價資料

針對商品的詢價、報價等主題的信件，是雙方商務合作關係中，最常需要使用的主題。

> **範 例**
>
> As requested in your letter of table lights, we are glad to enclose a copy of our new catalogue in which you will find many items that will interest you.
>
> 關於貴公司信件中對於桌上型燈具的詢價，隨信附上一份我方新的目錄，您可以找到有興趣的商品。

二、段落分明

和中文寫作原則相同，段落分明的e-mail書信，能幫助收信者在最短的時間內，就清楚地辨別您所要傳達的訊息。千萬不要毫不考慮段落，便將所有的訊息全部連接在一段文字中，這樣的書寫格式，會讓對方感到厭煩的。

> **範 例**
>
> In a recent issue of the "American Trade" from TIME magazine, we saw your name listed as being interested in making certain purchases in Taiwan.
>
> We take this opportunity to place our name before you as being a buying, shipping and forward agent.
>
> We've been engaged in this business for the past 20 years. We, therefore, feel that because of our past years' experience, we are well qualified to take care of your interest.

We look forward to receiving your reply in ac-knowledgement of this letter.

【中文翻譯】

我們在「時代雜誌」最新一篇「美國商業」的報導中，看見貴公司名列在「有意在台灣採購商品」的名單中。

希望藉由這個機會，向你介紹敝公司有關採購、船務及轉運的代理業務。

敝公司從事這個產業已經有廿年的經驗。有鑑於過去的經驗，我方是具有極佳的勝任能力來處理貴公司的權益。

期望能收到貴公司有關於此事的回覆。

錯誤範例

In a recent issue of the "American Trade" from TIME magazine, we saw your name listed as being interested in making certain purchases in Taiwan. We take this opportunity to place our name before you as being a buying, shipping and forward agent. We've been engaged in this business for the past 20 years. We, therefore, feel that because of our past years' experience, we are well qualified to take care of your interest. We look forward to receiving your reply in acknowledgement of this letter.

上述這種不分段落的書寫格式，不但閱讀不易，也顯得不尊重對方閱讀時的感受，如此一來，恐怕是會令收信者想要直接將e-mail刪除吧！

三、條列說明

除了要段落分明之外，若是提供的資料非常繁瑣，建議使用條列的方式來呈現，以方便對方閱讀或是確認。

> **範 例**
>
> The quotation is as follow:
>
> (1) Floor lamp: US$2,000 each, CIF Taipei
>
> (2) Table lamp: US$1,800 each, CIF Taipei
>
> 報價如下：
>
> (1) 立燈：美金2,000元/台　CIF台北
>
> (2) 桌燈：美金1,800元/台　CIF台北

四、簡潔的陳述

不論是正式或是非正式的書寫文體，都一定要強調「簡潔的陳述」方式，意思就是說，當您在書寫e-mail時，不要拖泥帶水，要用簡短、簡潔的方式，讓對方在最短的時間內，就能掌握您所發的這封e-mail的重點，亦即，無須太多過於冗長、修飾性的客套內容。

範 例

We have received your letter of July 2.
我方已經收到貴公司七月二日的信件。

傳遞訊息

1.七月二日

2.已收到信件

範 例

We appreciate your inquiry of October 25 about our goods.
感謝貴公司十月廿五日對我方商品的詢價。

傳遞訊息

1.十月廿五日

2.產品報價

範 例

Enclosed is our Agency Contract No. 1124 in detail.
請詳見附件編號 1124 號我方的代理權合約明細。

傳遞訊息

1.有附件可以參考

2.代理合約

範 例

We're very sorry for the delay for your inquiry on September 25.
很抱歉延誤貴公司於九月廿五日所提之詢價。

傳遞訊息

1.抱歉

2.延遲報價

五、簡單的用語

除了使用簡單的用語之外，也可以多多利用縮寫的格式來呈現，也能夠加強溝通的效率，例如：

(一) 主詞和動詞的縮寫

I am → I'm

he is → he's

she is → she's

we are → we're

they are → they're

I have → I've

I would → I'd

he would → he'd

she would → she'd

we would → we'd

(二) 否定式的縮寫

will not → won't

do not → don't

does not → doesn't

did not → didn't

should not → shouldn't

can not → can't/cannot

has not → hasn't

have not → haven't

(三) be動詞否定式的縮寫

is not → isn't

are not → aren't

was not → wasn't

were not → weren't

Part 3 加強溝通的效率

　　主旨列的撰寫是個不小的挑戰，因為只能用幾個簡單的字詞來陳述目的，而不能誤導收信者。

　　有一些人在寫e-mail時，容易犯了一些e-mail寫作上的毛病而不自知，往往容易造成溝通的負面效果，以下分析幾種可能書寫的主旨內容，藉此檢討不同的書寫內容，所傳達的意思對這一封e-mail的影響有哪些。

　　1. 主旨：｜　　　　　　　　　　　　　　｜（一片空白）

　　若是您自己收到一封主旨空白的e-mail，您會馬上迫不亟待就開啟這一封e-mail嗎？相信和大部分人的想法都是一樣的：若這不是一封夾帶病毒的e-mail，就是一封廣告的垃圾郵件，總之不會是一封重要的e-mail。這封e-mail的命運應該就是進入垃圾桶吧！

　　2. 主旨：｜Important! Read Immediately!!｜

　　這是一封標示「重要，馬上閱讀」的主旨列。試想，若真的重要，何不就將這個重要訊息節錄成為主旨呢？實在是

令人費疑猜！對發信者來說是重要的訊息，對收信者來說未必是同等重要，這樣的主旨內容完全無法達到吸引人的效果。

3. 主旨：| Quick question! |

「只是問一個問題」？這樣的主旨說明就和上述重要事件的道理相同，如果這是個非問不可的問題，就應該利用主旨列來提問。這種主旨說明是對e-mail的內容完全沒有幫助的主旨。

4. 主旨：| Follow-up about Friday |

這個主旨內容稍微好一些，能夠提供收信者記住這是一封要追蹤某一件在週五發生過的事，但是不夠完美，因為無法說明有哪些事需要收信者持續追蹤關心。

5. 主旨：| The file you requested |

雖然有說明是收信者要求的檔案，但是沒有說明是哪個檔案，所以無法讓收信者立即判斷是哪一個檔案，若是能改寫為："Sales file from David"則更為清楚，是由David所提供的"sales file"的e-mail。

除非您能非常確認，這封e-mail是收信者殷殷期盼的信

件，並能夠在收到信後就知道是您所寄發的，否則有可能是夾帶病毒的spam，因爲許多垃圾郵件都是使用這樣語意不詳的主旨說明，恐怕收信者還沒有看到信，就被收信者的郵件軟體判斷爲spam，而自動歸類到垃圾桶中吧！

6.主旨： 10 confirmed for Friday... will we need a larger room?

以上的主旨內容是個很棒的主旨說明，怎麼說呢？因爲點出了以下的重點：

（1）事件：十件確定的事

（2）日期：星期五

（3）問題：是否需要較大的房間

簡單、明確的主旨，只簡單地使用了10個字詞，就點出這封e-mail的重要議題，讓收信者可以立即針對是否需要"a larger room"的提問，考慮相關的問題。

撰寫主旨列的原則

一則合適的主旨說明，將直接影響您所發的這一封e-mail的效能。主旨列除了代表這封e-mail的溝通目的之外，也必須注意是否吸引收信者注意進而閱讀，而不會被淹沒在

眾多的e-mail中。以下的方法教您撰寫主旨列的一些基本原則，以提高您發出去的e-mail的閱讀率：

一、簡短、溫和

盡量保持主旨列的字詞在20個單字以內（包含空白鍵字數）。根據網路信件禮儀組織的最新研究報告指出，20個單字以內的主旨列較容易會被對方所開啟閱讀。所以若是想要以e-mail完成溝通傳輸的工作，請保持主旨列用詞簡短而溫和吧！

二、明確

主旨語意不清、不知所云甚至沒有重點說明的撰寫內容，都是無用的e-mail。通常的狀況常發生在電子報或例行的月報表中，例如："The Green Thumb Newsletter: June 2009"，這個訊息透露出的隱喻是「這封e-mail沒有什麼重要的消息需要宣告，不用打開來閱讀也沒關係」，比較適當、貼切的撰寫方式是"The Green Thumb: 3 Tips for Summer Gardening"（Green Thumb消息：三種夏日植栽的技巧），上述主旨簡單地陳述了三件事：

（1）The Green Thumb：發信者的身份

（2）3 Tips：提供一些小技巧的訊息

（3）Summer Gardening：主題內容是和「夏日植栽」
有關

三、主旨列留在最後撰寫

先完成e-mail的內容架構，等寫完整封e-mail後，再來
審視這封e-mail的內容，確認有哪些重點表達是足以當成主
旨列使用。

四、花時間思考主旨

不要只是匆匆忙忙寫完主旨列，要知道，主旨列是何等
重要的地位。可以先草稿三至四則主旨列，再從這些主旨列
中挑選合適的主旨來使用。

增進商務e-mail溝通效率

以下根據不同的主題，提供建議說明，以增進商務
e-mail溝通的效率。

一、主旨

根據調查顯示，一般人收到e-mail後，開信閱讀的習慣
動作是非常類似的，通常收信者都會先快速地瀏覽主旨列
後，再決定要先看哪一些他認為重要的e-mail信件，當然他
也有可能是決定將信件轉寄(forward)或刪除(delete)。請記

住，您的這一封e-mail信件並不會是收信者信箱中唯一的一封信。在您寄出e-mail信件前，好好地思考，哪一種主旨列可以提高您的e-mail信件的閱讀率。

爲了提昇雙方溝通的效率，減少誤解產生，e-mil的內容就必須把握「快、狠、準」的原則，因此e-mail的書寫就必須不拖泥帶水。首先必須考慮的，就是將「重要訊息」推上第一線。而這個第一線的重要地位，就非主旨欄位莫屬了。

商務e-mail的第一個重要的書寫重點呈現，就是在主旨的內容上。

主旨列是使收信者對你所發的e-mail產生興趣的重要功臣，也具有代表這一封e-mail的提點作用。因此，書寫e-mail時，主旨至少應該符合下列的目標之一：

（一）包含「重點訊息」：也就是達到訊息傳遞的作用。目的是期望對方開啓這封e-mail時，能夠立即得到的第一個訊息。

範　例

Sub: Sales meeting rescheduled to 2 P.M. on Friday
銷售會議改至週五的下午兩點鐘（召開）

傳遞訊息

1.會議改期

2.時間改至週五下午兩點鐘

範　例

Sub: Can you defrag my C drive?
　　可以幫我重整C磁碟機嗎？

傳遞訊息

1. C磁碟機需要重整

2. 請求幫助

（二）包含「期待的回應」：若是希望對方能針對你所發的這封e-mail的目的有所回應，你也可以將此「期待」（希望對方回應）明白告知對方。這種格式是要告知對方：「相關事件的說明在e-mail正文中，看完後請記得要回覆我」。

範　例

Sub: Your comments urgently needed by 4 P.M. today
　　請在今天下午四點鐘前提出您的建議

傳遞訊息
1. 請提出建議
2. 今天下午四點鐘前是最後期限

（三）明確而簡短：主旨列的說明必須簡短而明確，切忌冗長、沒有重點。請記住，時間有限、主旨列空間也很寶貴，不必浪費時間詳細說明，沒有人有那個美國時間花時間慢慢看你的主旨列的，所以請用簡潔的語句陳述主旨吧！

範　例

Sub: How about lunch tomorrow?
　　要不要明天中午一起吃飯？

傳遞訊息
1. 邀請共進午餐
2. 時間就在明天

範　例

Sub: Emergency: All Cars in the Lower Lot Will Be Towed in 1 Hour
　　緊急事件：所有在低樓層停車場的車子，將在一個小時後被拖吊

傳遞訊息
1. 低樓層車子
2. 一個小時後開始拖吊

（四）方便存檔：主旨列的說明也提供收信者瞭解，若是他必須將您提供的訊息或檔案存檔，那麼這一封e-mail就擔負說明此功能。

範 例

Sub: John's report
　　約翰的報告

傳遞訊息
1. 提供約翰的報告

要特別提醒你，若是您的這封e-mail並非單一訊息主題，而是有許多的訊息要傳遞，那麼建議您等完成整封e-mail後，最後再思考主旨列的內容。

因為您可以藉由審視整封e-mail後，再來決定哪一個重點必須要特別在主旨列中凸顯，也或許這個主旨列的內容，

是足以代表整封e-mail的主體架構。

二、提供發信者的聯絡電話

雖然說e-mail的溝通模式已為大部分人所接受並認同的溝通管道，可是還是免不了會發生需要用電話溝通的可能性。

盡可能在e-mail的署名處，註明您的聯絡電話，以方便萬一收信者需要與你電話聯絡時，可以不用再翻電話簿找出您的電話號碼，而可以在看e-mail的同時就撥打您的電話，幫助對方節省找電話的時間。也可以避免收信者剛好沒有您的電話的窘境。

範 例

Dear Sirs,

As requested in your letter of table lights, we are glad to enclose a copy of our new catalogue in which you will find many items that will interest you.

All details are shown in our price list. We are ready to deliver any quantity of table lights from stock.

Jack Smith

Sales Manager, Forever Books Company

(02)8647-3663

【中文翻譯】

敬啟者：

關於您信件中桌上型燈具的詢價，隨信附上一份我們新的目錄，你可以找到你有興趣的商品。

所有的細節都在我們的價格表中。我方準備好可以從現貨中運送任何數量的商品。

永續圖書公司銷售經理

傑克·史密斯 敬上

(02)8647-3663

在上述例句中，在傑克署名的下方提供的聯絡電話，讓收信者可以在電腦前閱讀信件時，萬一需要馬上與傑克聯絡，就可以直接撥電話，而不必再花時間找出傑克的電話。

而除了最重要的署名及聯絡電話之外，也有所代表的企業的名稱，方便收信者馬上知道是哪一家公司的「傑克」。

其他和署名相關可以提供身份的書寫內容。（詳見P59 說明）

三、刪除不必要的訊息

當收到一封e-mail時，發信者（sender）必然會有一些客套性的問候語，您回覆的時候，除了可以選擇附上原文的方式，也可以只保留和您的回覆相關的原文段落即可，因為如此一來，也可以節省對方閱讀e-mail所花費的時間。

範　例

發信者原文

>I hear you're working on the Smith account. If you need any information don't hesitate to contact with me.

回覆文

Dear Tom,

Listen, we've been working on the Smith account and I was wondering if you could give me a hand? I need some inside information on recent developments over there. Do you think you could pass on any information you might have?

Thanks

Peter

【中文翻譯】

>聽說你在處理Smith的帳戶。若是你需要幫忙，請通知我。

親愛的湯姆：

是這樣的，我們已經在處理Smith的帳戶，我在想，你是否能幫助我？我需要一些你們最新發展的內部消息。你是否有這方面消息可以提供給我？

多謝啦！

彼得

在上述的範例中，彼得應該是將對方的e-mail其中的一段原文"I hear you're …… contact with me"附加在他的回覆信件中，方便對方開啟e-mail閱讀時，知道傑克是針對這段原文所做的回覆。

四、善用顏色設定

除了上述將原發信者的原文附在回覆文中之外，另一種可以加強溝通效率的方式，就是附上對方的原文，並用顏色加以區別，可以讓收信者對您的回答一目了然，這種情形特別適用在對方的e-mail來信中，是用條列式方式提出希望你

回答或解決時使用。

範 例

>fee

$ 100 per person

>quantity

Two boxes

>date

Apr. 2nd

【中文翻譯】

>費用

每人100美元

>數量

2箱

>日期

四月二日

　　以上述範例為例，可以將原文與您的回覆用不同顏色標示，例如原文為黑色，回覆說明則為藍色，如此一來，針對來信者的問題，您的逐一回答便顯得有條不紊、清清楚楚。

五、同時CC給副本收件者

既然e-mail的發信者已經將此封e-mail同時CC給相關的收信者（CC的用法詳見P32說明），表示這是一件公開的事件，有點類似昭告天下的作法，那麼你回信（reply）時，也請選擇「全部回覆」的選項，亦即你有義務將此回覆內容，也同時寄送給發信者希望能一併通知的對象。

六、多此一舉的書寫格式

雖然說e-mail是為了提昇溝通的效率，但不代表可以無限上綱地詳細或是繁瑣，以下有一些書寫格式是不必在商務e-mail中出現的：

（一）無須電子郵件地址：不必刻意在e-mail中註明你的電子郵件地址（e-mail address），因為對方若是需要回覆（reply），只要按下「回覆按鈕」即可，你不必再多此一舉加註自己的e-mail address。但是若是你希望對方能利用你的另一個e-mail address回覆，則你可以在e-mail中特別註明。

範　例

You can reply me at my private email:
susan1988@yahoo.tw

你可以回覆到我的私人帳號中：susan1988@yahoo.tw

> **範　例**
>
> Please reply me at susan1988@yahoo.tw
> 請回覆到susan1988@yahoo.tw給我

　　（二）省略職稱：當對方回覆您的前一封e-mail，而你也需要再回覆給對方時，在e-mail最後的署名處，只要註明你的名字即可，亦即公司名稱、職稱等，都是屬於往返頻繁的e-mail中，不必一直重複出現的內容。

　　（三）禮多則怪：同樣地，在同一事件的頻繁往返中，客套問候或是寒暄、祝福等，就不必再費心思書寫了，這些客套的問候語句，在彼此發的第一封e-mail中就該具備，之後持續性溝通的e-mail，就可以省略這些客套用法了。

Part 4 商務實例

介紹公司

I have the pleasure of introducing our company to you.
有此榮幸向您介紹敝公司。

It's my pleasure of introducing IBM to you.
有此榮幸向貴公司介紹IBM。

We have the pleasure of introducing ourselves to you.
我方有此榮幸向您介紹敝公司。

We have the pleasure of introducing ourselves to you as one of the most reputable light exporters.
我方有這個榮幸向您介紹，敝公司是一家信譽優良的燈具出口商。

單 字	introduce 動詞 介紹
句 型	have the pleasure of ＋動/名詞
範 例	May I have the pleasure of this dance?
	能有這個榮幸邀您共舞嗎？

公司創立的時間

Established in 1990, we've been expanding our business operations around the world as a leading exporter and importer of cars.

在一九九〇年創立，我方已於全球拓展汽車的進出口業務。

Since our establishment in 1995, we have been marketing various electronic products in the USA.

自從一九九五年成立以來，我方已經在美國經營多種的電器商品。

We have been in the art business since 1995 and have never had an unhappy customer.

自從一九九五年起，敝公司便已經營此產業，從未有不滿意的客戶。

| 單　字 | establish | 動詞 | 建立、創立 |

句　型 since… ＋ 完成式（have/has+過去分詞）

範　例 We have finished the report since last Monday.
我們上週一就已經完成報告了。

片　語 in ＋ 西元時間

範　例 I was born in 1976.
我是一九七六年出生的。

多年經驗

We have been engaged in this business for the past 10 years.

敝公司從事這個業務已經有十年的經驗。

We have been specializing in your commodities for over 20 years.

本公司專營貴公司的商品已達廿年以上的經驗。

單 字	commodity 名詞 日用品
片 語	for ＋ 時間
範 例	for 10 days 長達十天的時間
	for past 2 years 過去長達十年的時間
慣用語	in this business 在此產業

公司的特色

B&B Company is a reliable company with wide and varied experience in the line in Taiwan.

B&B公司為一家在台灣信譽可靠的公司，在此產業擁有豐富的經驗。

B&B Company is the world's largest oil painting company with thousands upon thousands of the most satisfied and happy customers.

B&B公司是全球最大的油畫公司，擁有非常多滿意的客戶群。

We have been very successful in selling high quality artwork in the United States and Worldwide.

我方在全美及全世界成功地經營高品質的藝術商品。

We're an old and well-established exporter of all kinds of lights.

敝公司是老字號、制度健全、出口多種燈具的公司。

單　字	reputation　*名詞*　信譽	
	customer　*名詞*　客戶	
	successful　*形容詞*　成功的	
	importer　*名詞*　進口商　↔ exporter　出口商	
慣用語	a reliable company　值得信賴的公司	
	a well-established exporter　制度健全的出口商	
	high quality　高品質	
片　語	thousands upon thousands　成千上萬	
範　例	We have thousands upon thousands of collector cards for sale.	
	我們有成千上萬張卡片待售。	

說明產業

We take this opportunity to place our name before you as being a buying, shipping agent.

希望藉此機會，向您介紹有關敝公司採購、船務的業務代理。

Our main lines of products include: TV, Radio, MP3, CD player, etc.

我方的主要商品線包含：電視、收音機、MP3、CD播放器…等。

We are one of the leading exporters of cotton and rayon goods and are enjoying an excellent reputation through fifty year's business experience.

我方為純棉及尼龍製商品的出口商第一品牌,在過去五十年間的商務經驗,擁有良好的聲譽。

單 字	agent 名詞　代理商
	product 名詞　商品
慣用語	take this opportunity　藉此機會
	main lines of products　主要的商品線
	the leading exporters　具領導地位的出口商

拓展業務

As your name was listed in The Garden Magazine, we're writing in the hope of opening an account with your company.

因為貴公司名列在花園雜誌上,我方謹此來函,希望與貴公司開啟貿易合作。

We'd like to open an account with you, and hope that a mutually agreeable consideration of interests can be arranged.

我方渴望能與貴公司開啟貿易合作,也希望能建立雙贏的合作關係。

We're desirous of extending our connections in your country.
敝公司擬拓展本公司在貴國的業務。

單字	mutually 　*副詞*　 彼此、互相
慣用語	open an account　開始往來
範 例	We'd like to open an account with you. 我方渴望能與貴公司開啟貿易合作。
慣用語	extend our connections　拓展我方的業務
範 例	We're desirous of extending our connections in the USA. 我方擬拓展本公司在美國的業務。
片 語	be desirous of ＋ 動詞/名詞　亟欲做某事
範 例	We're desirous of extending our services. 我方擬拓展本公司的服務業務。

要求合作

At present we're interested in your goods.
目前我方對貴公司的商品極感興趣。

We're very interested in your lights.
我方對貴公司的燈具非常有興趣。

Now we're desirous of enlarging our trade in staple commodities, but have had no good connections in the USA.
目前敝公司亟欲擴大我方主要商品的業務範圍，但是我方目前在美國並無良好的合作關係。

If you are interested in our proposal, please let us know the terms and conditions of your business.

如果貴公司對我方的計畫案有興趣，請讓我方知道貴公司的合作方式及付款條件。

單 字	enlarge 動詞 擴大
	goods 名詞 商品
	staple 名詞 主要商品
	connection 名詞 關係
句 型	be interested in ＋ something　對某事有興趣
範 例	We're interested in your opinions!
	我們對你的想法很有興趣。

合作關係

We are one of the largest importers of woolen goods in Taiwan and shall be pleased to establish business relationships with you.

敝公司為台灣最大的毛線織品進口商之一，很高興和貴公司建立商務關係。

We're sure that you will be satisfied with our services and the excellent quality of our goods.

我方確信，貴公司會對我方所提供的服務及高品質的商品感到滿意。

These are very liberal terms and we feel your best interests are considered in this offering.

這些都是非常有利的條件，我方在此次報價中已將貴公司利益考慮在內。

單 字	quality 名詞 品質
	term 名詞 條件
	interests 名詞 利益
慣用語	establish business relationships 建立商務關係
範 例	We'd be delighted to establish business relationships with buyers worldwide.
	我方很高興能和世界各地的買家建立商務關係。
慣用語	be satisfied with ＋ something 對～感到滿意
範 例	I'm sure you'd be satisfied with our products.
	我確認貴公司會對我方的商品感到滿意。

回覆合作要求

Thank you for your letter of May 4 proposing to establish business relations between our two firms.

感謝您五月四日來函欲和我方建立雙邊商務關係。

單 字	propose 動詞 提議、提案
	firm 名詞 公司、企業
慣用語	thank you for ＋ something 因為某事感謝你

> 範例 Thank you for your promptly & professional service.
> 感謝您迅速而專業的服務。

否決合作計畫

We regret to inform you that we are not in a position to enter into business relations with any firms in your country because we have already had an agency arrangement with B&B Trading Co., Ltd. in Japan.

必須遺憾地通知您，因為我方已和日本的B&B商務有限公司有代理協議，所以我方無法和任何在貴國的公司有商務關係。根據協議，我方惟有透過上述公司才得以出口我方商品至日本。

We regret to inform you that we made our decision prior to receiving your proposal.

很遺憾地通知貴公司，我方在收到貴公司提案之前，就已經做出決定了！

> 單 字 regret 　動詞　 遺憾
>
> 　　　　inform 　動詞　 通知
>
> 慣用語 enter into　開始、加入
>
> 範 例 We have entered into a contract with NBC Company.
> 我們已經和NBC公司簽訂了合約。

慣用語	make someone's decision 某人已經決定
範 例	I have made my decision.
	我已經決定了。

信用查詢

B&B Co., who has recently proposed to do business with us, has referred us to your bank.
B&B公司最近要求和我方能有商業往來，對方指示我方向貴銀行查詢。

If you need more objective information concerning our credit, please refer to The Citi Bank.
如果貴公司需要更多我方的信用資料，請逕行向花旗銀行查詢。

Concerning our financial status and reputation, please direct all enquiries to The Citi Bank.
需要相關的財務狀況及信譽資料，請逕行向花旗銀行查詢。

We're pleased to inform you that we have completed the required credit inquiries, and the favorable results have justified our decision to enter into a business agreement with you.
我方已經完成必要的信用調查，有力的結果證明，我方決定與貴公司締結交易合作是有利的。

單 字　refer 動詞　查閱

　　　direct 動詞　徵詢

　　　credit inquiries　信用調查

慣用語　do business 生意往來

範 例　It's my pleasure to do business with you.

　　　很高興能和貴公司做生意。

慣用語　refer to 查閱、參照

範 例　He referred to the dictionary for the correct spelling
of the word.

　　　他查詢字典找出這字的正確拼法。

引薦客戶

We shall be obliged if you kindly introduce us to some
of reliable importers.

貴公司如能將我方介紹給一些信譽良好的進口商，我方將不
勝感激。

Therefore we shall be obliged if you could kindly introduce us to some of reliable importers in the USA who
are interested in these lines of goods.

因此若貴公司能將敝公司介紹給一些在美國對我方業務有興
趣、信譽良好的公司，我方將不勝感激。

單 字　obliged 形容詞　感激的

　　　kindly 副詞　好心地

> 慣用語 be obliged　感謝
> 範 例 I'd be obliged if you can help me with this question.
> 如果你能幫我解決問題，我將感激不盡

提供服務

We're confident to give our customers the fullest satisfaction.

我方有信心能令顧客感到十足的滿意。

Because of our past years' experience, we are well qualified to take care of your interest.

由於過去的經驗，我方具有極佳的勝任能力來負責貴公司的權益。

With our sufficient capital and intimate knowledge of business, we are sure that we can help you establish a successful business corporation in our area.

因為我方充分的資金及對此產業的瞭解，我方確信能幫助貴公司，在此地區建立成功的合作關係。

> 單 字 capital *名詞*　資金
> intimate *形容詞*　熟識的
> 慣用語 knowledge of business　產經知識
> 慣用語 be qualified　資格符合

> | 範 例 | She was qualified to teach high school mathematics and physics.
> 她夠資格教高中的數學和物理。
> | 片 語 | take care 處理
> | 範 例 | Let me take care of your business.
> 由我來處理你的事業。

代理權談判

We're very interested in your advertisement of May 20 seeking an agent here.

我方對貴公司在五月廿日刊登尋求本地的代理商的廣告極感興趣。

We recommend ourselves to act as your sole agent for your camera in Taiwan.

我方自薦能成為貴公司相機在台灣的獨家代理商。

Because you are not directly represented in Taiwan, we'd like to offer to be your agent.

因為貴公司在台灣還沒有直接的代理人，我方欲提供擔任貴公司在台灣的代理商業務。

With our long experience in marketing, we'd like to offer our services as your agent for Taipei.

因為在此產業有多年的經驗，我方希望能成為貴公司在台北的代理商。

We'd like to recommend our company as a most suitable agent for your products.

我方自薦，成為貴公司商品最適合的代理商。

Thank you very much for your proposal of May 25 offering us your services as New York Agents for our products.

感謝貴公司五月廿五日提案，希望能成為我方在紐約地區的代理商。

Thank you for your letter of May 25, in which you show your interest in being our supplier.

謝謝貴公司五月廿五日來函，表示有興趣成為敝公司的供應商。

We appreciate your request to act as our agent in your country, but we think it is premature for us to enter into agency agreement at the present stage.

感謝貴公司要求成為我方在貴國的代理商，但我方認為現階段簽訂代理協議尚為時過早。

單 字	advertisement 名詞 廣告
	seek 動詞 尋找
	recommend 動詞 推薦
	sole agent 獨家代理
	represent 動詞 作為～的代表

> offer 　動詞　 提供
>
> suitable 　形容詞　 適合的
>
> proposal 　名詞　 企畫案
>
> supplier 　名詞　 供應商
>
> appreciate 　動詞　 感謝
>
> premature 　形容詞　 過早的
>
> 慣用語 act as　扮演、充當
>
> 範　例　Mr. Smith acted as chairman in David's absence.
>
> 大衛不在時，史密斯先生充當主席。

簽署合約

Enclosed is our Agency Contract No. 278 in duplicate, a copy of which please sign and return to us for our file.

茲附上我方代理權編號278號合約，一式兩份，請署簽其中一份，並將它寄回我方存檔。

Please check all the terms listed in the contract and see if there is anything not in conformity with the terms we agreed on.

請您確認合約的所有條款，是否與協議的條款有不相符之處。

Can you speed up the contract and let us have it next Friday?

貴公司能否盡快在下星期五就完成簽訂合約的時程？

We have already signed a contract for a period of five years with another supplier in Taiwan.

我方已經與另一家在台灣的供應商簽訂了五年的合約。

單 字	contract 名詞 合約
	check 動詞 確認

慣用語 speed up 加速

範 例 We developed a new system to help speed up the work.

我們研發一套新的系統以協助加速工作效率。

慣用語 sign a contract 簽訂合約

範 例 Before signing a contract, you should check all the terms.

簽訂合約前，你應該要確認所有的條款。

市場需求

We're in a position to handle large quantities.

我方目前有大量的訂單需求。

We're interested to purchase canned pineapple in large quantity.

我方極有興趣採購大量的鳳梨罐頭。

單 字	quantity 名詞 數量
	purchase 動詞 購買

| 片 語 | in a position　　在某個立場 |
| 範 例 | We're in the position to ship our goods by this Friday.
我方是站在週五前出貨的立場。 |

要求提供商品型錄

Will you please send us a copy of your catalogue?

貴公司是否能寄給我方一份型錄？

Will you please send us a copy of catalogue, with details of the prices and terms of payment?

請寄給我方一份型錄，並註明價格和付款條件。

單 字	catalogue　名詞　型錄
句 型	send ＋ someone ＋ something　寄某物給某人
範 例	Please send me a copy of your catalogue. 請寄給我方一份貴公司的型錄。
句 型	Will you please ＋原形動詞　能否請你～
範 例	Will you please send us a copy of catalogue? 能否請您寄一份型錄給我方？

要求提供樣品

We should find it most helpful if you could also supply samples of your goods.

如果貴公司能提供商品的樣品，將對敝公司有極大的幫助。

If available, please send us a piece of sample.

如果方便，請寄給我方一件樣品。

Can you mail us some free samples?

您能寄免費樣品給我方嗎？

May I request some samples of your products before placing a formal purchase order?

我方能在正式下單前，要求提供商品的樣品嗎？

單 字	supply	動詞	提供
	sample	名詞	樣品
	request	動詞	提出要求
慣用語	if available, ... 如果方便的話，～		
範 例	If available, please send me your offer.		
	如果方便的話，能否提供報價。		

提供樣品

We already sent you some free samples yesterday.

我方已於昨日寄給貴公司一些免費樣品。

We're glad to send you samples of our MP3 you inquired.

謹此寄予貴公司所要求我方提供的MP3樣品。

As requested in your letter dated May 25, we sent you the samples via UPS on May 28.

貴公司五月廿五日來函索取的樣品，我方已於五月廿八日以 UPS寄出。

We're sending you some free samples. Please give us your specific inquiries upon examination of the above as we presume they will be received favorably in your market.

隨信附上一些免費樣品，在檢視過以上的樣品後，請告知貴公司的特別需求，我方相信，此樣品必能符合貴公司的市場需求。

We're truly sorry that we can't send you samples.

非常抱歉，我方無法提供樣品。

We can't send free samples.

我方無法提供免費樣品。

You may place a sample order.

您可以下樣品的訂單。

慣用語	as requested　依照要求
範　例	As requested, we sent you a piece of sample via UPS yesterday.
	依照貴公司的要求，我方昨天以UPS寄出一件樣品給貴公司。
慣用語	letter dated ＋日期　某個日期所發的信

範 例 Thank you for your letter dated Oct. 10. We've sent you 2 boxes of sample.

感謝您十月十日的來函。我方已經寄給貴公司兩箱的樣品。

慣用語 place a sample order　下樣品訂單

範 例 Should we place a sample order?

我們應該要下樣品的訂單嗎？

收到樣品

Thanks for the latest catalogue and 20 pairs of running shoes received last Tuesday.

感謝貴公司，我方已於上週二收到貴公司所寄的最新的目錄和 20 雙運動鞋！

Thank you for the samples, which you sent to us on Sep. 10.

謝謝貴公司九月十日寄給我方的樣品。

We have received your samples last week.

我方上週已收到樣品。

We have received the samples that you sent to us last week.

我方已收到貴公司上週寄出的樣品。

單　字　receive　動詞　接收

句　型　on ＋月份日期　在某個日期

範　例　My birthday is on May 30th.

　　　　我的生日是在五月卅日。

慣用語　thanks for something　因某事感謝

範　例　Thanks for your sample.

　　　　多謝貴公司提供的樣品。

慣用語　thank you for something 因某事感謝

範　例　Thank you for calling.

　　　　多謝來電。

詢價

We would like to make an inquiry.

我們想要詢價。

Please quote for us the prices of the items listed on the enclosed inquiry form.

請依所附需求表，為我方報價。

We have many inquiries for the under-mentioned goods.

我們收到許多下述貨品的詢價單。

We are interested in your products. Please quote for us.

我方對貴公司的商品有興趣。煩請提供報價。

Please quote for us the latest prices for five tables.

請提供給我方五張桌子的最新報價。

Please send us your best CIF quotation for lamps.

請提報我方燈具的最優惠的CIF價格。

Please advise your offers, minimum quantity and conditions.

請提供貴公司的報價、最小訂購量及付款條件。

Would you kindly please quote for us for your best offer with CIF Taipei?

能請貴公司提供CIF台北的最優惠報價嗎？

Will you please send us a copy of your catalogue, with details of prices and items of payment?

貴公司是否能寄給我方一份貴公司的型錄，並附上價格、付款明細？

We shall be greatly obliged if you will send us a copy of your catalogue, informing us of your best terms and lowest prices CIF Taiwan.

假使貴公司能寄貴公司的型錄給我方，並告知貴公司銷售最佳商品CIF到台灣的最低價格，我方將不勝感激。

單　字　inquiry form　需求表

inquiry　*名詞*　詢價

quote　*動詞*　報價

CIF quotation　*名詞*　CIF 報價

advise　*動詞*　告知、建議

offer　*名詞*　報價

minimum quantity　最小訂購量

condition　*名詞*　付款條件

慣用語　make an inquiry　詢價

範　例　We'd like to make an inquiry of your Model 2598.
我方想要對型號2598商品詢價。

提供報價

We would love to make an offer about sporting T-shirts.
我方很願意對運動T恤提供報價。

We hope you will be satisfied with our samples and quotations.
我方希望貴公司能對我們的樣品和報價感到滿意！

Please let us know as soon as possible if our offer does not contain what you want in order to send you another new quotation.
如果我方提供的價格不符合貴公司的期望，煩請盡速告知，以便另外提供新的報價。

This is a combined offer on all or none basis.

此為聯合報價，必須全部接受或全部不接受。

All prices are subject to change without notice.

所有的報價隨時更動，恕不另行通知。

We renew our offer of April 10th on the same terms and conditions.

我方基於原來條件，更新四月十日之報價。

We're very sorry for the delay for your inquiry on October 25.

很抱歉延誤貴公司於十月廿五日所提之詢價。

單　字	terms and conditions　付款條件
	renew　動詞　更新
	delay　名詞　延遲
句　型	be satisfied with something　對某事感到滿意
範　例	We're not satisfied with your quotations.
	我方對貴公司的報價不甚滿意。
慣用語	make an offer 提供報價
範　例	We have made an offer to you last week.
	我方已於上週提出報價給貴公司。
慣用語	as soon as possible 儘早
範　例	Please call me back as soon as possible.
	請盡早回我電話。

片 語	in order to ＋ 動詞　為了做某事
範 例	In order to visit my parents, I have to be home at 7 PM. 為了要拜訪我的父母，我需要在晚上七點鐘到家。

提供報價資料

We enclosed a copy of our price list.

隨信附上一份我方的價格清單。

Enclosed is our Prices List No. 200906-245.

附件為我方編號 200906-245 之價格明細清單。

We thank you for your enquiry of September 25. Enclosed please find our price quotation.

感謝貴公司九月廿五日的詢價。附件為我方的報價明細。

As requested in your letter of table lights, we are glad to enclose a copy of our new catalogue in which you will find many items that will interest you.

貴公司來函詢價桌上型燈具，隨信附上一份我方新的目錄，僅供參考。

單　字	enclose　動詞　附上
	price list　報價表
	enquiry　名詞　詢問

> **句 型** a copy of ＋ something　一份文件
> **範 例** Will you send us a copy of your latest price list?
> 請寄給我方一份貴公司最新的價格表。

對價格不滿意

We won't accept your quotation.

我方將無法接受您的報價。

We regret that we are not in a position to accept the order at the prices.

很抱歉，我方無法按此價格接受訂單。

I'm afraid your price is above our limit.

您的價格恐怕超出了我們的界限。

We'll place a trial order with you if the prices are competitive.

如果價格具競爭性，我們將會下試驗性訂單。

I don't think we would make a bad purchase.

我不認為我們會用高價購買。

The price you offered is out of line with the market, so it is beyond what is acceptable to us.

您的報價與市場情況不相符，故我方無法接受。

單字 accept 動詞 接受

at the prices 依此價格

limit 名詞 限制

competitive 形容詞 具競爭力

beyond 副詞 遠不及

acceptable 形容詞 可被接受的

慣用語 be acceptable to someone　對某人來說是可接受的

範例 I'm afraid the price you offered is not acceptable to us.
您所提供的價格，恐怕我方無法接受。

議價

If you can make the prices a little easier, we shall probably be able to place an order.
如果能再降一點價格，我方也許會下訂單。

Is it possible to shade the prices a little?
有沒有可能降一點價格？

The Japanese quotation is much lower.
日本的報價就比較低。

The quotations we received from other sources are much lower.
我方從別處得到的報價要低得多。

| 單字 | possible | 形容詞 | 可能的 |

單字　possible　形容詞　可能的
　　　shade　動詞　略減（價格等）
　　　lower　形容詞　較低的←→higher　較高的
句型　be able to + 動詞　能做某事
範例　I won't be able to offer you the quotation on time.
　　　我將無法如期報價。
慣用語　be much lower 非常低的
範例　The quotation is much lower than you expected.
　　　報價比你原先預期的還要低。

下訂單

We are going to make an order.
我方打算下訂單。

We are going to place an order.
我方即將下訂單。

We've settled the terms for the contract in general, and we are going to place an order.
大體上我方已將合約條款都談妥了，接著我方就要下訂單了。

We are ready to place an order with you, but only one condition is that goods are confined to Taiwan.
我準備向貴公司下訂單，但是唯一的條件是，貨物只限賣給台灣的公司。

The following products have been selected from your price list; please supply them to us immediately.

從貴公司的價格表中，我方已選購以下的商品，請立即提供商品給我方。

單 字	settle 動詞 安排
	condition 名詞 條件
	immediately 副詞 立即地
句 型	make an order 下訂單 = place an order
範 例	We recommend you to place an order as soon as possible.
	我方建議貴公司盡早下單。
慣用語	be ready to ＋動詞 準備好要做某事
範 例	I'm not ready to send you the samples.
	我還沒有準備好要寄樣品給你。
片 語	be confined to 在某地受到限制
範 例	These goods may be confined to the USA.
	這些只限制在美國地區。
片 語	in general 一般而言
範 例	In general, we'd place an order on Friday.
	通常我們會在週五下訂單。

取消訂單

We regret that we have to cancel our order.

很抱歉，我方不得不取消訂單。

We regret that we have to cancel our order because of the inferior quality of your products.

很抱歉，由於貴公司商品品質低劣，我方不得不取消訂單。

For all consequences arising from cancellation, we consider that you are all held liable.

我方認為貴公司對於取消訂單所引起的一切後果，應負全責。

單　字	cancel 動詞　取消
	cancellation 名詞　取消
	inferior 形容詞　不良的
	consequences 名詞　後果、影響
	consider 動詞　考慮
句　型	because of ＋名詞
範　例	We cancel the game because of the rain.
	因為下雨，我們取消比賽。

接受訂單

Thank you for your order.

謝謝貴公司下訂單。

Thank you for your Order No. 1130-1.

謝謝貴公司編號 1130-1 號的訂單。

We're pleased to receive your Order No. 1130-1 and confirm acceptance of it.

很高興接到貴公司編號1130-1號的訂單，並確認予以接受。

Please notify us of the purchase number.

請告知我方訂單的編號。

We shall fulfill your order by November 30.

我方會在十一月卅日前完成貴公司的訂單需求。

We're ready to supply any quantity of your order.

我方準備好提供貴公司訂單所需之數量。

Thank you for your order. We accept it and will dispatch the goods in the early of August.

謝謝貴公司下的訂單。我方接受此訂單，並將於八月初交貨。

單 字　order　名詞　訂單

confirm　動詞　確認

acceptance　名詞　接受

notify　動詞　通知

purchase number　訂單編號

fulfill　動詞　完成

dispatch　動詞　迅速處理、發送

句 型　by ＋時間 在某個時間點之前

> **範 例** I'll place an order by this Friday.
> 我會在本週五前下訂單。

催促下訂單

We would advise your order without loss of time.
我方建議貴公司應該不要猶豫,立即訂購。

As we are booking heavy orders, we would advise your order without loss of time.
因為我方的訂單量非常大,我方建議貴公司毫不猶豫地立即下訂單。

Once the order is confirmed, we need you to give us a pre-advice two weeks ahead you placing an order for we need to prepare the materials.
一旦訂單確認後,請在下訂單的前兩週預先通知我方,因為我方需要準備生產物料。

單 字	without	副詞	沒有、無
	as	連接詞	因為
	once	連接詞	一旦
	ahead	副詞	預先、事先
	prepare	動詞	備妥
	material	名詞	原料
慣用語	without loss of time		立即地、毫不遲疑地

> **範 例** Try to place an order without loss of time.
>
> 請嘗試著立即下單。

出貨

Please inform us of the shipping date.

請通知我方裝船的日期。

Please ship the enclosed order immediately.

請立即安排所附訂單的出貨事宜。

Please tell us when our order will be shipped.

請告知我方的訂單何時會裝船。

I'd appreciate it very much if you could deliver the goods by this Friday.

如果你們能在這個星期五前出貨，敝公司將不勝感激。

As the market is sluggish, please postpone the shipment of the order No. 203 goods to August.

由於市場疲軟，請將敝公司訂單編號 203 延遲至八月出貨。

Many thanks for your letter of August 15, in which you inform us of the date of shipment.

感謝貴公司八月十五日來信通知我方船運的日期。

單 字	shipping date	出貨日期
	ship 動詞	出貨
	deliver 動詞	運送
	shipment 名詞	出貨
	sluggish 形容詞	不景氣的
	postpone 動詞	延遲
片 語	inform someone of something	通知某人某事
範 例	I informed Mr. Smith of David's safe arrival.
	我通知史密斯先生，大衛已然平安抵達。

通知出貨

We're pleased to inform you that your order has been shipped.

謹此通知，貴公司訂單已經出貨了。

As the goods you ordered are now in stock, we'll ship them without fail immediately.

因為貴公司訂購的商品尚有存貨，敝公司將立即安排出貨。

Your order for 100 doz. hats will be shipped at the end of this month. You will receive them early next month.

貴公司一百打帽子的訂單會在這個月底裝船出貨。您將會在下個月初收到這批貨。

單 字	in stock　庫存
	early　*副詞*　提早地↔late　晚地
句 型	without fail　必定、一定
範 例	We'll ship your goods without fail.
	我們會立即出貨。

要求提早交貨

I need to discuss the date of shipment with you.
我需要和您討論船運出貨的時間。

When is the earliest date you can ship your goods?
貴公司最早什麼時候可以出貨？

Would you please reschedule shipment to the middle of September instead?
能請您重新安排船運的日期至九月中旬嗎？

Could you ship the first consignment by mid September?
貴公司能在九月中旬交第一批貨嗎？

Shipment has to be made before September; otherwise we're not able to catch the season.
九月份以前貨必須裝載上船，否則就趕不上銷售季節了。

I'm afraid that shipment by the middle of September will be too late for us.
九月中旬才能交貨對我方來說恐怕會太晚。

單字	discuss	動詞	討論

單字 discuss 動詞 討論

earliest 最早地 副詞 ↔ latest 最晚地

reschedule 動詞 重新安排時程

middle 名詞 中間

instead 副詞 替代

consignment 名詞 運送、托運物

otherwise 副詞 否則

片語 the middle of ＋月份 某月的中旬

範例 We'll ship your goods by the middle of September.
我方會在九月中旬前出貨。

交貨時程

We will dispatch the goods in the early of August.
我方將於八月初交貨。

The earliest shipment we can make is the middle of April.
四月中旬是我方能夠安排的最早的出貨時間。

I'm very sorry, but we really can't advance the time of delivery.
非常抱歉，但是我方真的無法提前交貨。

I am sure that the shipment will be made not later than the beginning of September.
我可以保證，交貨期不會遲於九月上旬。

In spite of our effort, we find it impossible to secure space for the shipment owing to the shortage of shipping space.

雖然敝公司已盡最大努力，卻因為船位不足而無法保證交期。

I'm sorry to say that the delivery is a week behind schedule.

很抱歉，交貨時間比計劃的行程晚了一個星期。

單 字 advance 動詞 提早

shortage 名詞 不足、短缺

delivery 名詞 運送

impossible 形容詞 不可能的

慣用語 in the early of ＋月份在某個月份的上旬

範 例 We'll be ready in the early of October.

我方會在十月上旬就準備好。

片 語 in spite of ＋ something 不管、不顧某事

範 例 In spite of numerous failures, David finally succeeded.

儘管多次失敗，大衛最終仍舊成功了。

片 語 behind schedule 比預定時間晚

範 例 It's five days behind schedule.

比預定時間晚五天。

商品品質不良

Our customers complain that the goods are much inferior in quality to the samples.
我方客戶抱怨貨物的品質遠低於樣品。

The products you supply are much inferior.
貴公司提供的商品品質非常低。

Your goods are much inferior in quality to my knowledge.
貴公司的商品品質低於我方的認知。

單 字	complain	動詞	報怨
	knowledge	名詞	認知、瞭解

對商品的抱怨

I'm afraid I've got a complaint about the quality.
恐怕我得要稍微抱怨關於品質的問題。

We regret to complain that your consignment of goods is not of the quality and color of the sample piece.
貴公司商品的品質和顏色與樣品不符，我方遺憾地對此表達不滿。

I'm writing to inform you that I am dissatisfied with your products.
謹此來函通知，我方對貴公司商品的不滿。

We have received your goods yesterday. But I'm sorry to say this quality is not so satisfactory.

我方已於昨日收到貴公司的商品。但是很遺憾，這個品質不太令人滿意。

I would say that it was unwise of you to supply such unqualified goods to us.

我只能說，貴公司提供不合格的商品給敝公司，是非常不明智的舉動。

I believe you shall adopt strict measures to control the quality of your own products.

我認為，貴公司要採取嚴格措施來控制商品的品質。

This is the first complaint we have ever received in our long business relation with you.

此為雙方長久合作以來，我方第一次提出抱怨。

We deeply hope you are fully committed to preventing this sort of problem.

我方衷心期望貴公司盡全力防止問題的發生。

| 單 字 | complaint | 名詞 | 抱怨 |

dissatisfied 形容詞 不滿意的↔satisfied 滿意的

satisfactory 名詞 滿意↔unsatisfactory 不滿意

unwise 形容詞 不明智的↔wise 明智的

unqualified 形容詞 品質不良的 ↔ qualified 品質合格的

adopt 動詞 採取

strict measures 嚴格的措施

control 動詞 控制

prevent 動詞 阻止、避免

problem 名詞 問題

慣用語 主詞 + deeply hope 某人誠摯地期望

範 例 We deeply hope you accept our proposal.
我方誠摯地期望貴公司接受我方的計畫。

句 型 commit to ＋名詞/動名詞 承擔、擔保

範 例 He didn't commit himself to anything.
他沒有作任何承諾。

回覆對商品的抱怨

Thank you for your letter of May 25 in which you claim the wrong color of the hat.
感謝貴公司五月廿五日信函所提，有關帽子顏色不對的申訴。

We regret your dissatisfaction with our product.
我方很遺憾貴公司不滿意我方的商品。

單 字 claim 動詞 要求、主張

dissatisfaction 名詞 不滿意

商品損毀

We're now returning the damaged items and would be grateful if you would replace them immediately.

如今我方將此損毀的商品退回，希望貴公司能盡快替換新品，我方將感激不盡。

We're sorry to learn that your goods were badly damaged during transit.

很遺憾得知，貴公司貨物在運送途中嚴重受損。

We're terribly sorry to learn from you that the quality of goods is not satisfactory to you.

我方為商品品質無法令貴公司滿意而感到抱歉。

We're very sorry to learn from your letter of 10th May that our shipment covering your order No. 569-4 was found defective.

很抱歉，由五月十日的信件中得知，我方的貨運所運送貴公司訂單編號 569-4 號商品的瑕疵。

We hope you will immediately propose a settlement to offset the damage.

希望貴公司針對此傷害，儘速提出解決方案。

單 字	return 動詞 退回
	damaged item 損毀商品
	replace 動詞 替換、替代
	learn 動詞 瞭解、認知、得知
	transit 名詞 運送
	defective 形容詞 有瑕疵的
	propose 動詞 提議
	settlement 名詞 解決
	offset 動詞 補償
片 語	be grateful 感謝
範 例	I am grateful to have you help me solve the problem.
	承蒙您幫忙解決問題，我方感激不盡。
慣用語	be terribly sorry 非常抱歉
範 例	We're terribly sorry for the damaged items.
	我方非常抱歉商品有損毀。

商品短缺

We have noticed a discrepancy between our invoice and the quantities you specified.

我方發現，敝公司的發票上的數量與貴公司所提的數量有不符之處。

I'm writing to inform you that the short delivery of our Model B423.

謹此來函通知，型號B423 的商品有短缺。

We're sorry for the short delivery.

我方為交貨量短缺向貴公司致歉。

單 字	notice 動詞 發現、注意
	discrepancy 名詞 不一致
	between 介係詞 在～之間
	invoice 名詞 發票
	specify 動詞 說明
	short delivery 商品短缺
片 語	discrepancy between/in＋某事 某事不一致、有差異
範 例	There was a discrepancy in the two minutes of the meeting.
	關於那場會議的兩份備忘錄有矛盾之處。
慣用語	be writing to inform someone 來函通知某人
範 例	I'm writing inform you that we found the shortage of Order No. 264-9.
	謹此來函通知，我方發現訂單編號 264-9 號商品的短缺。

賠償事宜

We would ask you to cover any loss.

我方將要求貴公司負責任何的賠償。

We hope you would compensate us for the loss.

希望貴公司賠償我方損失。

We are prepared to accept your claim.

我方準備接受貴公司的索賠要求。

We can't but lodge a claim against you.

我方不得不提出索賠。

As regards inferior quality of your goods, we claim a compensation of NT$200,000.

由於貴公司商品質量低劣，我方要求賠償廿萬新台幣。

單 字 cover 動詞 提供保險

loss 名詞 損失

compensate 動詞 賠償

compensation 名詞 賠償

句 型 lodge a claim 提出抗議

範 例 He lodged a complaint with the general manager.

他向總經理提出了抗議。

句 型 as regards 至於

範 例 As regards the loss, we will claim a compensation.

至於損失，我方會要求賠償。

接受賠償

In order to solve the problem, we ask for compensation for the loss.

為了解決問題，我方為所受的損失要求賠償。

The insurance company will compensate you for the losses according to the coverage arranged.
保險公司將按照投保險種賠償損失。

Any expense you may charge us will be paid by our company.
任何貴公司所要求的費用，將由我方支付。

Any expenses you may incur in this connection will be gladly paid upon being notified.
任何因此而衍生的費用，經正式通報後，我方將會願意支付。

單 字	insurance company 保險公司
	coverage 名詞 保險項目
	expense 名詞 費用
	charge 動詞 索取費用
	incur 動詞 發生
	connection 名詞 業務往來
句 型	according to 根據
範 例	We will go or we won't, according to circumstances. 我們或去或不去，都將視情況而定。
慣用語	solve the problem 解決問題
範 例	Do you have any idea how to solve the problem? 該如何解決問題，你有任何想法嗎？

付款條件

When sending you our samples, we also informed you of the price with samples.
寄樣品給您的同時，我方也已一併通知樣品的費用。

As to payment, we'll draw you a draft at 90 d/s an D/A.
至於付款，我方將以承兌交單的條件，開立見票九十天後付款。

Attached please find a sample letter of credit under which a draft may be drawn at sight.
附件為可以開立即期匯票的信用狀樣本。

Your last month's payment will be paid by remittance.
最後一個月的付款將會以匯寄的方式支付。

We won't accept payment in cash on delivery, but may consider payment in cash with order.
我公司不接受貨到付款的支付方式，但可以考慮隨單付款的支付方式。

The amount concerned was forwarded to your account of the Taipei Bank by telegraphic transfer today.
上述的金額已於今日電匯至貴公司台北銀行的帳號。

Our terms of payment are by irrevocable L/C payable by sight draft against presentation of shipping documents.

我方的支付方式是用不可撤銷的信用狀支付,憑裝運單據見票付款。

Payment by L/C will give the best protection to the exporters.

用信用狀付款為出口商提供了最好的保護。

Please extend the L/C to July 20.

請將信用狀的有效期限延至七月廿日。

We hope you will take action to assist us with our financial difficulties.

希望你方採取措施,幫助我們克服資金困難。

We will not accept L/C 45 days. Please change it to L/C at sight.

我方無法接受見票四十五天後付款的信用狀條款,請將之修改為即期信用狀。

Please see the enclosed copy of our wire payment.

請參考我方附件的電匯繳款影本。

單 字 d/s=days after sight 見票後～天付款

D/A=documents against acceptance 承兌後交單據

at sight　即期

remittance　*名詞*　匯款

in cash on delivery　貨到付款

in cash with order　隨單付款

amount　*名詞*　金額

telegraphic transfer　電匯

L/C　信用狀

protection　*名詞*　保護

慣用語 take action 採取行動

範　例 We have to take action to stop them.
我們得採取行動來制止他們。

催款

We request your immediate payment.
我方要求貴公司立即付款。

Up to this date, it appears that we haven't received your remittance.
截至目前為止，顯示出我方尚未收到貴公司的匯款額。

This is to inform you that payment on our invoice No. 4216-2 is now more than 45 days overdue.
現通知貴方 4216-2 號貨運清單的付款已超期 45 天。

As you know, the due date for payment was May 8, but as of today we have not yet received your remittance.

如您所知，付款的到期日為五月八日，但截至今日，我方仍未收到貴方的匯款。

We have received no response from you concerning our request for payment dated May 18. Please let us know if there is a problem.

我方要求貴方五月十八日付款，但至今未有回覆，請告知是否有問題。

Our records indicated that NT$8,000 had not been paid since May. We would say that it was unwise of you to have done that.

根據我方的記錄，貴公司新台幣八千元的費用自從五月起就尚未結清。我方認為貴公司所為極度不明智。

We therefore ask you to settle the account immediately.

因此請貴方立即支付。

If found in order, please favor us with your remittance within the period stipulated.

若查核無誤，請在規定時間內賜予匯款。

According to the terms and conditions of the monthly account agreed upon, we would like to receive settlement by 20th May.

依照每月清單同意的合作條款，本公司願於五月廿日前收到清償款項。

We're wondering why your remittance of payment for 100 dozen umbrellas does not reach us despite our repeated requests.

令人不解，不顧我方的持續催款，貴公司仍未支付一百打雨傘的匯款。

Please let us have your remittance immediately. If not, we will be obliged to take legal steps.

請立即支付匯款。假如無法支付款項，我方將採取法律行動。

We shall appreciate it if you will give us an another fort-night to settle your account due September 15.

如果貴公司能再給予十四天的付款寬限至九月十五日，我方將感激不盡。

We resend a photocopy of our invoice No. 4561-2 dated June 25.

再寄去一份我方六月廿五日 4561-2 號貨運清單的影本。

單 字	immediate	形容詞	立即的
	appear	動詞	顯示
	overdue	形容詞	過期的
	response	名詞	回覆、未兌的
	indicate	動詞	顯示

> the period stipulated 規定的時間內
>
> despite 介係詞 儘管
>
> 片語 up to this date 截至目前為止
>
> 範例 We haven't received your payment up to this date.
>
> 截至目前為止，我方尚未收到貴公司的付款。

解釋尚未付款

We're terribly sorry for the delay in payment.

我方對逾期付款深感抱歉。

We apologize for the delay of payment.

對於逾期付款，我方深感抱歉。

We have received your letter, dated today, by fax and apologize for the delay in payment.

我方已收到您今日的傳真，很抱歉我方付款延期。

We have enclosed a copy of appliance sheet of remittance, please check it with your bank.

內附我方向銀行提交的匯款申請書的影本一份，請與貴公司的銀行核對。

I don't understand why you haven't received our remittance until now.

我不明白為何貴公司至今未收到我方的匯款。

| 單 字 | apologize | 動詞 | 致歉 |

慣用法 until ＋時段　到～為止

範 例 Will you finish this report until tomorrow?
你到要明天才會完成報告嗎？

即將付款

Due to our internal remittance procedures, which took longer than anticipated, we could only make our remittance application to our bank today.
由於本公司內部匯款手續比預計時間要長，我方只能今天向銀行提出匯款申請。

You should be able to confirm the receipt of the funds in your account on May 5, your time.
貴公司應在當地時間五月五日收到匯款。

We'll correct the mistake. I'm sure the remittance should be arrived tomorrow.
我方馬上糾正了錯誤。我保證匯款明天就會到。

We apologize for the delay payment, and we will make the remittance in two days.
很抱歉逾期付款，我方將在兩日內付款。

Please extend one week for the payment limit.
懇請再予以寬限一周付款。

We corrected the situation immediately and you should be able to receive the funds tomorrow.

我方馬上糾正了錯誤，貴公司明天就會收到款項。

We apologize for the delay and we will make the payment within the next two days.

很抱歉逾期付款，我方將在兩日內付款。

單 字	internal 形容詞 內部的
	procedure 名詞 手續
	receipt 名詞 收據
	fund 名詞 現款
	account 名詞 帳戶
片 語	due to ＋名詞 由於～原因
範 例	Her absence was due to the storm.
	由於風雨交加，所以她沒來。
慣用語	correct the mistake 糾正錯誤
範 例	I promise we'd correct the mistake immediately.
	我保證我方會立即改進錯誤。

已收到款項

If you have already made payment, please disregard this notice.

如貴方已付款，請不必理會本通知。

We have received your remittance of US$1000.

我方已經收到貴公司美金 1000 元的匯款。

We have received your remittance in settlement of our claim.

我方已經收到貴公司解決我們索賠問題的匯款。

After we've received your remittance, we'll e-mail you for your confirmation.

待收到貴公司的匯款，我方將會以電子郵件寄發確認函。

單　字　disregard 　動詞　 不予理會
　　　　notice 　名詞　 通知
　　　　confirmation 　名詞　 確認函
慣用語　make payment　 付款
範　例　I'm sorry that we forgot to make payment.
　　　　很抱歉，我方忘記付款了。

Part 5 禮貌性用語

感謝來函

We have received your letter of July 25.

我方已經收到您七月廿五日的信件。

Thank you very much for your proposal of May 25.

非常感謝貴公司五月廿五日的提案。

We have received your letter of March 3 about the quotation.

我方已經收到您三月三日有關報價的信件。

I take great pleasure in receiving your offer letter dated July 10, 2009.

我很高興收到您二〇〇九年七月十日的報價。

單　字	thank 動詞 感謝
	letter 名詞 信件
	pleasure 名詞 高興、愉悅
慣用語	Thank you very much for + something
	因某事感謝您
範　例	Thank you very much for your support.
	多謝您的大力協助。

> **慣用語** take great pleasure in + 動/名詞　因某事感到非常高興
>
> **範　例** I take great pleasure in introducing B&B Company to you.
> 很高興在此向您介紹B&B公司。

來信目的

I write to express myself.
謹此來函表達我自己的想法。

I write to explain things.
謹此來函解釋說明。

I'm writing to you to ask your opinion.
謹此來函請教您的意見。

I'm writing to let you know that we are going to meet Mr. Jones on Monday.
謹此來函讓您知道，我們星期一要和瓊斯先生見面。

I'm writing to inform you that your sales project is canceled.
謹來信通知您，您的銷售計畫已取消。

I'm writing to tell you about your products.
謹此來函告訴您，有關貴公司商品一事。

We write to ask if you are interested in our products.
謹此來函詢問，貴公司是否對我們的產品有興趣。

單字　write　[動詞]　書寫

express　[動詞]　表達

explain　[動詞]　解釋

ask　[動詞]　詢問

opinion　[名詞]　意見、想法

meet　[動詞]　會面

project　[名詞]　企畫案

慣用語　let someone know　使某人知道

範例　If you have any question, please let me know.
如果您有任何問題，煩請告知。

來函祝賀

I'm writing to congratulate you for the wonderful work you all have been doing.
謹此函恭喜您有如此棒的工作表現。

I'm writing to congratulate you all for the great job you are doing.
謹此來函恭喜您的優良成就。

I'm writing to congratulate you and your committee on this successful experiment.
謹此來函恭喜您及委員們這一次成功的試驗。

I'm writing to congratulate you both on a job well done.
謹此來函恭喜二位工作表現良好。

單　字	congratulate　*動詞*
	wonderful　*形容詞*　極佳的
	committee　*名詞*　委員會
	successful　*形容詞*　成功的↔unsuccessful　不成功的
	experiment　*名詞*　試驗
	both　*副詞*　兩者
片　語	congratulate someone for something　恭賀某人某事
範　例	I'm writing to congratulate you for your success. 謹此來函恭賀您的成功。

來函回覆

I'm writing in reply to your letter asking for advice.
謹此來函回覆您來信要求協助一事。

I'm writing in reply to your letter of September 14, in which you accuse me of unauthorized use.
謹此來函回覆您九月十四日的來信有關質疑我的非授權性使用。

I'm writing in reply to your letter dated 15 May 2009.
謹此來函回覆您二〇〇九年五月十五日的來信。

I'm writing in response to John's letter.

謹此來函回覆約翰的來信。

單 字	accuse 動詞 質疑
	unauthorized use 未被授權使用
片 語	in reply 做為答覆
範 例	I'm writing in reply to give you some suggestion.
	謹此來函回覆予您一些建議。

敬請回覆

I'd appreciate having your reply.

能有您的回覆,我將感激不盡。

I'd appreciate having your reply by the 1st of April.

您能在四月一日前回覆,我方將不甚感激。

Your replies will be appreciated about those issues.

感謝您關於此議題的回覆。

Your special attention to this will be highly appreciated.

承蒙特別關照此事,我方將不甚感激。

Please answer promptly.

請立即回覆。

We're looking forward to your prompt reply.

期待您的立即回覆。

We're looking forward to hearing from you soon.

期待盡快得到您的回覆。

We look forward to receiving your reply in acknowledgement of this letter.

我方期望能收到您有關於此事的消息。

We should be pleased if you would respond to our request at your earliest convenience.

如果您能儘早回覆，我方將不甚感激。

單　字	issues 名詞　議題
	special 形容詞　特別的、特殊的
	attention 名詞　注意
	promptly 副詞　快速地
	soon 副詞　迅速地
	acknowledgement 名詞　確認
	convenience 名詞　方便
慣用語	be highly appreciated　非常感謝
範　例	I'd be highly appreciated if you could reply to my new proposal soon.
	如果您能針對我新的提案儘速回覆，我方將不甚感激。
片　語	look forward to ＋ 名詞/動名詞　期望某事
範　例	I'm look forward to your earliest response.
	期待您的盡速回覆。

範　例　I'm looking forward to hearing from you soon.
期望盡快得到您的消息。

片　語　in acknowledgement of　承認

範　例　I bowed my head in acknowledgement of guilt.
我低下頭來承認過失。

歡迎提問

Please call me any time if you have any questions.
如果有任何問題，歡迎您隨時打電話給我。

Please feel free to contact us if you have any questions.
若您有任何問題，請與我方連絡。

Feel free to contact me.
歡迎與我方聯絡。

If you have any further questions, please feel free to contact us.
如果您有更進一步的問題，歡迎與我方聯絡。

Please call me at 886-02-86473663 if you have any questions.
如果您有任何的問題，請打電話到 886-02-86473663 給我。

單　字　call　動詞　打電話
question　名詞　問題

> contact 動詞 聯絡
>
> further 形容詞 進一步的、更深入的
>
> 慣用語 have any questions 有任何問題
>
> 範例 Just let me know if you have any question.
> 如果有任何問題，告訴我一聲。
>
> 慣用語 feel free to contact 隨時可以聯絡
>
> 範例 Please feel free to contact us.
> 請儘管和我方聯絡。

通知

I inform you that we have removed to more convenient premises, situated at the above address.
謹此通知，我方已遷到上述更方便的地址。

We are pleased to inform you that our business will be turned into a limited company on 1st May.
謹此通知，本公司於五月一日將改為股份有限公司。

I am not sure about it. Can I inform you later this week?
這件事我不確定，我可以本週晚一點通知您嗎？

Please send us an e-mail when finding the report.
當您找到報告時，請用電子郵件通知我方。

We would like to inform you that we are not accepting the committees.

謹此通知您，我方不會接受委員會職務。

This is to inform you that Mr. Jones will arrive in New York tomorrow night.
謹此通知，瓊斯先生將會在明天晚上抵達紐約。

I am happy to inform you that your application is a great idea.
謹此通知，您的申請案是一個很棒的主意。

In reply to your letter, we are pleased to inform you that we have shown the sample to our buyer.
謹此回覆，我方已將樣品提交本公司的買方，特此奉告。

| 單 字 | premises | 名詞 | 辦公室場所 |

situate 動詞 位於

report 名詞 報告

arrive 動詞 抵達

idea 名詞 想法、主意

片 語 arrive in 抵達某地

範 例 I arrived in New York five days ago.
我五天前抵達紐約。

慣用語 tomorrow night 明天晚上

範 例 Will you come to the meeting tomorrow night?
你明天晚上會來參加會議嗎？

客套語

Thank you for your inquiry.

感謝詢價。

Let me begin by thanking you for keeping me informed of your latest catalogue.

首先感謝您通知我有關貴公司的最新型錄。

I take great pleasure in receiving your inquiry letter dated May 10, 2009 on the running shoes.

我很高興收到您二○○九年五月十日有關運動鞋的詢價信函。

Thank you for your interest in our products.

感謝貴公司對敝公司產品的興趣。

I'd like you to confirm this appointment by return.

很樂意您能回覆信函確認此次的約會。

We appreciate your early confirmation.

我方很感激您的儘速確認。

I do appreciate the effort you're making towards concluding this transaction.

感謝您為達成這筆交易所做的努力。

單 字	confirmation 名詞 確認
	effort 名詞 努力
	conclude 動詞 締結、推定
	transaction 名詞 交易、買賣、業務
片 語	keep someone informed 通知某人
範 例	Please keep me informed via e-mail
	請用e-mail通知我。
慣用語	do ＋原形動詞 的確有做某事
範 例	I do finish the report by myself.
	我的確是自己完成這份報告的。

道歉

I'm really sorry about that.
我真的為那件事感到很抱歉。

I owe you an apology for my rudeness yesterday.
我為昨天的粗暴行為向你道歉。

I apologize to you for not attending the meeting yesterday.
我因為沒有出席昨天的會議而向您表示歉意。

I regret to say that I am unable to help you.
很抱歉，我愛莫能助。

We still feel sorry for the trouble that has caused you so much inconvenience.

我方仍然感到抱歉，對您造成這麼多不便的困擾。

We're terribly sorry that we made so much trouble to you.
很抱歉，我方對貴公司製造了這麼多的麻煩。

Please accept our deepest apology for any inconvenience this matter has caused you.
請接受敝公司對此事所造成貴公司之不便的深深歉意。

I feel sorry that I am not able to make a corresponding move.
我為自己不能做相應的行動深感遺憾。

We apologize for not replying to you earlier.
很抱歉未能盡早回信。

We have no excuses, only apologize the inconvenience for you.
我方沒有任何藉口，只能為造成貴方的不便表示歉意。

單 字	sorry	形容詞	抱歉
	owe	動詞	欠
	apology	名詞	道歉
	rudeness	名詞	無理
	apologize	動詞	道歉
	trouble	名詞	麻煩

cause　*動詞*　引起

inconvenience　*名詞*　不方便↔convenience　方便

matter　*名詞*　事件

corresponding　*形容詞*　相同的、符合的

慣用語 attend the meeting　參加會議

範　例 Did you attend the sales meeting yesterday?

你昨天有參加會議嗎？

慣用語 make so much trouble　製造如此多的麻煩

範　例 I'm sorry David made so much trouble to you.

很抱歉，大衛給你闖了這麼多禍。

慣用語 accept someone's deepest apology　接受某人深深的

歉意

範　例 Please accept my deepest apology.

謹此致上我深深的歉意。

安撫

We hope this unfortunate incident will not affect the relationship between us.

我方希望這一不幸事件，將不會影響雙方之間的關係。

Thank you for your patience.

感謝您的耐心。

Thank you for the trouble you've taken in this matter.

謝謝您為這件事費心了。

We solicit your close cooperation with us in this matter.
我方懇求您對於此一事件能給予協助和合作。

We are confident to give our customers the fullest satisfaction.
我方有信心能提供給客戶最完整的滿意。

We hope to be of service to you and look forward to your comments.
以上希望能對您有所幫助，也靜待您的指教(意見)。

Please understand that this is really an exception, and we'll do our utmost to correct it.
請諒解此為特例事件，我方將盡力改進。

We have taken immediate action to make the necessary arrangements.
我方已採取緊急措施，做了必要的安排。

We have to request your waiting.
我方懇求貴公司的等待。

Please accept our sincere apologies for the delay.
因為此一延誤，請接受我方真誠的道歉。

單　字 unfortunate *形容詞* 不幸的 ↔ fortunate 幸運的
incident *名詞* 事件

affect 動詞 影響

relationship 名詞 關係

patience 名詞 耐心

solicit 動詞 懇求

confident 形容詞 有信心的

comment 名詞 見解、建議

exception 名詞 例外

correct 動詞 更正

necessary 名詞 需要

arrangement 名詞 安排

sincere 形容詞 忠誠的

delay 名詞/動詞 延遲

慣用語 do someone's utmost 盡某人的全力

範例 We'll do our utmost to avoid the accident.
我方會盡力避免意外。

催促

The above inquiry was forwarded to you on January 15, but we haven't received your reply until now. Your early offer will be highly appreciated.
上述詢價已於一月十五日行文給貴公司，可是我方到現在還沒收到貴方答覆，請早日報價，不甚感謝。

We haven't received your quotation until the date of May 10th.
我方直到五月十日都還未收到您的報價單。

單　字	above　*形容詞*　上述的
	until　*副詞*　直到某個時間點
慣用語	forward to someone　轉寄給某人
範　例	Would you please forward this letter to Mr. Smith?
	請你把這封信轉寄給史密斯先生好嗎？

靜待回音

Your prompt reply will be appreciated.

請盡速回覆，我方將不甚感激。

Your kind reply will be much appreciated.

貴公司的善意回覆，我方將不甚感激。

We look forward to the pleasure of hearing from you.

我方將期待貴公司的回音。

We're looking forward to your prompt confirmation on above request.

我方期待儘早收到貴公司針對以上要求所做出的回覆。

We recommend this matter to your prompt attention.

我方建議貴公司立即針對此事回覆。

Your courtesy will be appreciated, and we earnestly await your reply.

對於您的協助，我方將感激不盡，敝公司將靜待您的回音。

Thank you for your time, and we are looking forward to hearing your reply as soon as possible.

感謝撥冗，我方期待盡快得到貴公司的回覆。

單 字	attention	名詞	注意
	courtesy	名詞	禮儀、禮貌
	earnestly	副詞	真誠地
	await	動詞	等待

慣用語 hear from someone 得到某人的消息

範 例 I look forward to hearing from you as soon as possible.

我期待盡快得到您的消息。

片 語 as soon as possible 越快越好

範 例 Please reply us as soon as possible.

請盡快回覆我方。

提供進一步資訊

If you need any information, please contact us or visit our website.

如果有任何問題，歡迎與我方聯絡或瀏覽我們的網站。

We advise you to visit our website:

http://www.foreverbooks.com.tw

我方建議您瀏覽我方的網站：

http://www.foreverbooks.com.tw

If you have any questions, please do not hesitate to let me know.

如果有任何問題，請不要延遲，立即讓我知道。

If there is anything remaining unclear, you are always the most welcome to contact us.

如果有任何不清楚之處，歡迎隨時與我方聯絡。

Any information you may give us will help us to solve this problem.

任何您所提供的訊息，將有助於我們解決這個問題。

單　字	advise	動詞	提出建議

單　字　advise　動詞　提出建議

hesitate　動詞　遲疑

remain　動詞　保持

unclear　形容詞　不清楚的↔clear　清楚的

visit website　瀏覽網頁

慣用語　be always the most welcome to ＋動詞　歡迎做某事

範　例　You're always the most welcome to call us.
　　　　歡迎來電詢問。

片　語　solve the problem　解決問題

範　例　Do you have any idea how to solve the problem?
　　　　你有任何想法該怎麼解決問題嗎？

要求配合

We'll be in touch with you right after receiving your thought to the above questions.
得知貴公司對以上問題的看法後，我方將會與貴公司聯絡。

Thank you for your cooperation.
謝謝貴公司的合作！

Thank you for your arrangement.
謝謝貴公司的費心安排。

Thank you for the trouble you've taken in this matter.
謝謝貴公司的費心處理。

We hope that you will make every effort to avoid similar mistake in our future transactions.
希望貴公司會盡一切努力去避免未來的貿易關係，發生類似的錯誤。

Any help you may give us will be thankful.
貴公司所提供給的任何幫助，我方都將感激萬分。

單 字	thought	名詞	想法
	cooperation	名詞	合作
	in this matter		關於此事
	avoid	動詞	避免

similar 　*形容詞*　相似的

mistake 　*名詞*　錯誤

thankful 　*形容詞*　感謝的

慣用語 be in touch with someone　和某人保持聯絡

範　例 Try to be in touch with us.

請試著與我方保持聯絡。

慣用語 make every effort　盡一切努力

範　例 We'll make every effort to solve the problem.

我方會盡一切努力來解決這個問題。

合作關係

We look forward to the pleasure of working with you again in the very near future.

期待在不久的將來，能再與貴公司合作愉快。

We'd be happy to continue to cooperate in any way.

敝公司將盡全力與貴公司繼續合作。

Thanks for all the assistance that you have provided with so far.

感謝貴公司到目前為止提供予我方的各種協助！

We hope to have the pleasure of serving you again soon.

希望還有機會為貴公司服務。

單 字	continue	動詞	繼續、持續

cooperate 動詞 合作

assistance 名詞 協助

句 型 in the very near future 不久的未來

範 例 We'd compensate you for the shortage in the very near future.

因為短缺，我方很快就會而賠償貴公司。

片 語 provide with ＋ something 裝備、供給某物

範 例 The machine is provided with radar equipment.

這部機器裝裝置有雷達設備。

片 語 so far 截至目前為止

範 例 We haven't received your remittance so far.

截至目前為止，我方尚未收到您的匯款。

提出要求

I'm writing to request that the error should be corrected.
謹此來函要求更改錯誤。

I would ask you to bring it back to me.
我將要求您將它帶回來給我。

I would ask them not to tell anyone about it.
我們將要求他們不要告訴任何人這件事。

We would ask him to make this phone call.
我們將要求他打這通電話。

I would ask you to keep the secret.

我方將要求您守住秘密。

Your special attention to this will be highly appreciated.

承蒙特別關照此事，我方將不甚感激。

單　字	error 名詞 錯誤
慣用語	make a phone call　打電話
範　例	Why don't you just make a phone call to Mr. Smith?
	你何不打個電話給史密斯先生？
慣用語	keep the secret　保守秘密
範　例	Would you keep the secret for me?
	您可以為我守住秘密嗎？

Part 6 事務性洽談

建議

Here is what I propose...
以下是我的提議～

My proposal has two parts.
我的提議分為兩個部分。

I propose to begin tonight.
我建議今晚就開始。

We should be glad if you would help us with your suggestions.
如果您能提供您的建議，我方將會非常高興。

I advised against their doing it.
我勸他們不要做這件事。

He suggested that those entries be blanked out.
他建議刪去那幾個項目。

Thank you again for your proposal and your understanding of our position will be appreciated.
感謝您的建議，對於您能夠體諒我方的立場，我方不甚感激。

| 單　字 | propose | 動詞 | 提議 |

單　字 propose *動詞* 提議
part *名詞* 部分、一段
suggestion *名詞* 建議
understanding *名詞* 理解

詢問意見

What do you think?

您覺得呢？

What do you think about it?

您覺得這個如何？

Are you free next Tuesday at two o'clock?

下個星期二兩點鐘您有空嗎？

Do you agree with me?

您同意嗎？

單　字 think *動詞* 認為、想法
about *介係詞* 關於
片　語 agree with ＋ someone　同意某人
範　例 I totally agree with you.
我完全同意您。

介紹伙伴

I'm writing this e-mail for Mr. Smith who plans to be New York for about 2 weeks.

我是幫史密斯先生寫這封電子郵件，他打算去紐約兩週。

We have great pleasure in introducing to you, by this e-mail, Mr. Smith, president of IBM Company.

我方有此榮幸，藉由此封電子郵件，將IBM公司總裁史密斯先生介紹給您。

Mr. Smith is the manager of Sales & business development for Asia market.

史密斯先生是亞洲事業發展部的經理。

Mr. Smith is the president of IBM Company who has a long business experience in Medical and Technology.

史密斯先生是IBM公司總裁，他在醫療和科技業有長久的商務經驗。

Mr. Smith, our sales manager, will be spending 3 days in New York shortly on business.

史密斯先生是我方的業務經理，他打算在紐約短期出差三天。

Mr. Smith would welcome the opportunity of seeing you while he is in New York.

史密斯先生在紐約的時候，很希望能有機會和您見面。

單 字	plan	動詞	計畫
	development	名詞	發展
	spend	動詞	花費
	welcome	動詞	歡迎
片 語	on business		為了公事、為了生意
範 例	Mr. Smith will be in New York on business for 2 weeks.		
	史密斯先生會去紐約出差兩週。		

要求會面

Mr. Smith would like to meet you on the morning of 10th May.

史密斯先生想要在五月十日早上與您見面。

I think we should meet and discuss it a little more.

我想我們應該見面討論。

I'd like to make an appointment to see you.

我想要跟您約個時間見面。

I'd like to see you tomorrow if you have time.

如果您有空，我想明天跟您見個面。

I'd like to talk about it more often if you have time to-morrow.

假使您明天有空，我們可以再討論。

When can we meet to talk?

什麼時候我們能見面談一談？

Would it be possible for us to talk to Mr. Black in person about that?

我們可否親自跟布來克先生談論此事？

Could we meet and discuss the matter in a little more detail?

我們可以見面再詳細討論一下這件事嗎？

單字 discuss 動詞 討論

慣用語 make an appointment 約定會面

範例 Could you make an appointment with Mr. Smith for me?

可以幫我安排和史密斯先生會面嗎？

慣用語 have time 有時間

範例 Do you have time next Friday?

您下星期五有空嗎？

慣用語 in person 親自

範例 I need to talk to Mr. Smith in person.

我需要和史密斯先生親自談一談。

安排行程

I'll arrange this meeting.

由我來安排這次會議。

I suggest that we should meet at two pm.

我建議我們下午兩點鐘見面。

How about the day after tomorrow at five o'clock in the afternoon?

後天下午五點鐘如何？

How about tomorrow night at seven o'clock? Is it OK with you?

明天晚上七點鐘如何？您可以嗎？

Let's tentatively say next Monday at two o'clock.

我們暫時約定下星期一的兩點鐘。

單 字	arrange　*動詞*　安排	
	tentatively　*副詞*　暫時性地	
慣用語	how about ＋ 名詞/動名詞　針對～的意見如何？	
範 例	How about having a meeting with David?	
	要不要和大衛開會？	
慣用語	in the afternoon　在下午	
範 例	I'm writing to inform you that the meeting scheduled to two o'clock in the afternoon.	
	謹此來函通知，會議改期至下午兩點鐘。	
片 語	the day after tomorrow　後天	
範 例	I'd send you the sales report by the day after tomorrow.	
	我會在後天之前寄銷售報告給您。	

見面約定生變

I have to cancel tomorrow's appointment.
我必須取消明天的約會。

Something happened, so I have to fly to Hong Kong this afternoon.
發生一些事，所以我今天下午要搭飛機去香港。

I'm afraid Mr. Jones has to cancel your appointment.
恐怕瓊斯先生必須取消和您的約會。

John asked me to inform you that he is not going to visit you tomorrow.
約翰要我通知您，他明天無法去拜訪你了。

John said he is unable to be there.
約翰說他無法去了。

I'm afraid I'll have to postpone the appointment.
恐怕我必須把約會延期。

單 字	postpone 　動詞	延遲
慣用語	be unable to ＋ 動詞	無法做某事
範 例	I'm unable to be there on time.	
	我無法準時抵達。	
慣用語	be afraid	恐怕

> **範 例** I'm afraid we won't make it in time.
>
> 我們恐怕無法及時完成。

提供資料

The quotation is as follow:

(1) Notebook: US$2,000 each, CIF New York City

(2) PC: US$1,800 each, CIF New York City

報價如下：

(1) 桌上型電腦　美金 2,000 元/台 CIF 紐約市

(2) 個人電腦　　美金 1,800 元/台 CIF 紐約市

In order to give you an idea, we quote some of them without engagement as follows:

(1) Bags, cotton & fibers mixed

(2) White, US$20 per doz. CIF your port

為了讓貴公司有一些概念，我們主動提供一些非合約性商品的報價，如以下所列：

(1) 提袋，棉和纖維混紡

(2) 白色，美金 20 元/打，交貨條件：CIF 至貴公司指定的港口。

Please also indicate the delivery time in your quotation sheet.

請在報價單中註明交貨時間。

Click here to see some of our latest hand painted master-pieces.

請連結此處查詢我方的最新手繪作品。

Please let us know what you can offer in this line as well as your sales terms, such as mode of payment, time of delivery, discount, etc.

請告訴我方貴公司能提供什麼樣的商品，以及貴公司的銷售條件，如付款方式、交貨日期、折扣等。

If any further information is required, please let us know.

如果需要更進一步的資訊，請通知我方。

Please check all details.

請確認所有的細節。

Please check all the documents for me.

請為我檢查所有的文件。

If you need any information, please contact us or visit our website.

如果您有任何問題，歡迎您與我們聯絡或瀏覽我們的網站。

Please find the attached information and our best offer for our products.

請參考附件資料和我們所提供最優惠的商品價格。

單 字	indicate 動詞 指示、說明

indicate 動詞 指示、說明
delivery time 運送時間
quotation sheet 報價單
mode of payment 付款方式
discount 名詞 折扣
detail 名詞 細節
document 名詞 文件

溝通協調

Shall I suggest that we meet each other half way?
讓我們各作一半的讓步吧!

We should be glad to hear at your earliest convenience the terms and conditions on which you are prepared to supply.
得知您盡早準備提供商品,我方將不甚感激。

We would say that it was unwise of you to have done that.
我們只能說,您的所為是非常不明智的。

It seems to us that you ought to have done that.
似乎對我們而言,您應該做那件事。

We're afraid we cannot comply with your request.
我們恐怕無法順從您的要求。

We offer an opportunity of discussing a contract with you.
我們提供和您討論合約事宜的機會。

I'm so sure we can do something to help you.
我確定我們能作些事來幫助你。

The three sides reached an agreement to stop this project.
三方達成協議停止這項計畫案。

It's no problem for us to sign a contract with you.
對我們來說，和貴公司簽約不成問題。

If negotiation fails, it shall be settled by conciliation.
如果協商失敗，就需要調解。

單字	suggest	動詞	提出建議
	ought	助動詞	應當、理當
	negotiation	名詞	協調
	fail	動詞	失敗
	conciliation	名詞	安撫、調解

片語 comply with 遵守

範例 I'm sure we'd comply with all rules.
我確認我方會遵守全部的條例。

慣用語 meet each other half way 各退一步、各自退讓

範例 We could meet each other half way if you want.
如果你希望的話，我們可以各自退讓一步。

> | 慣用語 | reach an agreement　達成共識 |
> | 範　例 | I'm glad we have reached an agreement on the terms of the contract. |
> | | 很高興我們就合約條款已經達成共識。 |
> | 慣用語 | sign a contract　簽訂合約 |
> | 範　例 | Are you ready to sign the agency contract with us? |
> | | 貴公司準備好和我方簽訂代理權合約了嗎？ |

抱怨

We have to complain to you about the damage in shipment, which has caused us so much trouble.
我們不得不向貴公司抱怨，裝運的破損給我方造成很大的麻煩。

We regret to inform you that the goods shipped per S.S. "Beauty" arrived in such an unsatisfactory condition.
我們遺憾地告知，貴公司由「美女」號輪船運來的貨物令人十分不滿。

The importer has filed a complaint with our corporation about poor packing of the goods.
進口商為貨物的不良包裝向我公司提出抱怨。

We can assure you that such a thing will not happen again in future deliveries.
我方向貴公司保證，這樣的事件在以後的出貨中不會再發生。

We regret to find out the total shortage of your goods is 10 pieces.

我方相當遺憾地得知，貴公司商品的短缺總額是十件。

I regret to say that I am unable to help you.

很抱歉，愛莫能助。

I regret that you see it like that.

貴公司如此看待這件事，甚感遺憾。

單　字　assure　動詞　確認、保證

片　語　complain to someone　向某人抱怨

範　例　I'm afraid we have to complain to you about the shortage.

恐怕我方得向貴公司提出短缺的抱怨。

維繫合作關係

We are ready to be at your service and await your order.

我方已準備好為您服務也正等待您的訂單。

We have not received your order since last September.

自從去年九月起，我方就沒有接獲貴公司的訂單。

We thank you for your cooperation for our business for the past five years.

感謝貴公司過去五年來與我方的商務合作。

The matter has been given through consideration here.

敝公司已做過全盤的考量。

單 字	past *形容詞* 過去的
句 型	be ready to ＋ 動詞　準備好做某事
範 例	We're ready to accept your order.
	我方準備好接您的訂單。

確認達成共識

I'm glad that we've agreed on terms of the agency after the past four months.

很高興在過去四個月中，雙方就代理權一事取得共識。

片 語	agree on ＋ something　同意某事
範 例	I'd agree on your sales proposal.
	我會同意你的銷售計畫。

孩子我們一起學英語！(5-8歲)（附MP3）

為孩子創造最生活化的英語環境！
一對一互動式生活英語，讓孩子與您一起快樂學英語。英語學習絕非一蹴可幾，每天20分鐘的親子對話，孩子自然而然脫口說英語！

我的菜英文【單字篇】

英文開口說的音譯寶典。
全方位英文口語學習，不用高學歷也會開口說英語！不必再從頭學音標了利用中文發音學習英文快速克服發音的障礙！

我的菜英文【工作職場篇】（附MP3）

超強職場英語，一次收錄/警察/計程車司機/秘書/辦公室人員/銷售人員/飯店櫃臺。
英文開口說的音譯寶典。用中文音譯學習英文，不用高學歷也會開口說英語！讓您輕鬆工作上的說英語的需求！

超簡單的旅遊英語（附MP3）

易學、好用、最道地。
300組情境對話，海外自助旅遊者的溝通寶典，再也不必害怕面對外國人！主題情境×實用例句，使用超簡單的英語，也能全球走透透！

出差英文1000句型（附MP3）

出差英文寶典。明天就要出差了，該如何和外籍伙伴、客戶溝通？誰說英文不好就不能出差？只要帶著這一本，就可以讓你國外出差無往不利、攻無不克！

三合一職場英語（會議+商務+寫作）（附MP3）

缺一不可的英語實力：1.會議2.商務3.寫作多管其下，全方位提升職場的競爭力！Part 1會議加強參加會議時的表達能力Part 2商務主動出擊，提升談判溝通的能力Part 3寫作提點寫作的技巧，輕鬆完成商務寫作！

使用最頻繁的旅遊英語（附MP3）

超完整！出國旅遊必備寶典！ 從登機、出境、入境到食宿、旅遊，完全搞定！簡單、實用、完整，一定可以輕鬆上手的旅遊英語，讓你開開心心出國旅遊！

這本片語超好用

徹底擺脫英語學習的障礙！利用動詞片語串連式的學習法，快速掌握英語單字、生動的動詞片語應用，有助於增進開口說英語的技巧！

用最簡單的英文來回答

本書除了集結食、衣、住、行、育、樂基本口語，亦囊括了情緒表達、醫療衛生、電話及辦公室口語。一本書幫您行遍天下，用最簡單的英文句子，就可以清楚回答所有問題囉！

最簡單的英文自我介紹（附MP3）

最深刻的第一印象，來自最精采的自我介紹，利用短短二十秒，準確地推銷自己，更是增進人際關係的關鍵。讓我們用最簡單的句子，幫助您完成一篇最漂亮的自我介紹吧！

用你會的單字說英文（附MP3）

你絕對認識的單字，你絕對聽過的會話，但你就是不知道該何時使用、如何組合，每次講了兩個字就停住了，不知該如何說下去？？？EXCUSE ME……．BYE！BYE！只要會國中小基本1200單字，你也可以開口說英文，就算遇到外國人或出國也不怕了

你也能看懂運動英語

愛運動的你，可以經由本書將英語學習融入生活中。不懂運動的你，也可以從這裡開始入門。

你肯定會用到的500單字(50開)

背單字＝聽力UP！+寫作能力UP！單字不必多背，只要選對單字背，英文程度一定高人一倍！
基礎單字六大分類，讓您一次奠定英文學習基礎！Chapter 1基礎/Chapter 2能力/Chapter 3行為/Chapter 4身份/Chapter 5說明/Chapter 6時間

Good morning很生活的英語(50開)

超實用、超廣泛、超好記，好背、好學、生活化，最能讓你朗朗上口的英語。日常生活中，人們要透過互相問候來保持一種良好的社會關係。

超簡單的旅遊英語（附MP3）(50開)

出國再也不必比手劃腳，出國再也不怕鴨子聽雷，簡單一句話，勝過會背卻派不上用場的單字，適用於所有在國外旅遊的對話情境。

你肯定會用到的500句話(50開)

最實用的情境對話，出國必備常用短語集！
如果你以為學好英文一定要從背單字、念文法開始，那你就中了英文教科書只是學習法的蠱毒了！簡單情境、實用學習，用最生活化的方式學英文，簡單一句話，就能解決你燃眉之急！

商業祕書英文E-mail(50開)

商業秘書英文E-mail英文寫作速成班！我也可以是英文書寫高手！
明天就要寫一封信外國客戶了，卻不知道該如何下筆嗎？？本書提供完整的E-mail書信範例，讓您輕輕鬆鬆完成E-mail書信。

遊學必備1500句 (50開)

留學‧移民‧旅行，美國生活最常用的生活會話！遊學學生必備生活寶典，完全提升遊學過程中的語言能力，讓您順利完成遊學夢想！

菜英文-基礎實用篇 (50開)

沒有英文基礎發音就不能說英文嗎？不必怕！只要你會中文，一樣可以順口秀英文！

學英文一點都不難！生活常用語句輕鬆說！只要你開口跟著念，說英文不再是難事！

如何上英文網站購物 (50開)

上外國網站血拼、訂房、買機票…，英文網站購物一指通！

利用最短的時間，快速上網搜尋資料、提出問題、下單、結帳…從此之後，不必擔心害怕上外國網站購物！

出差英文1000句型 (附MP3) (50開)

出差英語例句寶典，完整度百分百！實用度百分百！簡單一句話，勝過千言萬語的說明！適用於所有國外出差的情境。

1000基礎實用單字 (附MP3) (50開)

只要掌握基礎單字，輕鬆開口說英語！蒐集使用最頻繁的1000單字，編撰最實用、最好記、最能朗朗上口的口語英文！

菜英文【旅遊實用篇】 (50開)

就算是說得一口的菜英文，也能出國自助旅行！本書提供超強的中文發音輔助，教您輕輕鬆鬆暢遊全球！

國家圖書館出版品預行編目資料

商務英文 E-mail ／張瑜凌編著.

--初版. --臺北縣汐止市：雅典文化, 民 98.08

　　面；公分. --（E-mail 系列：7）

　　ISBN：978-986-7041-83-8 (平裝)

　1. 商業書信　2. 商業英文　3. 商業應用文　4. 電子郵件

493.6　　　　　　　　　　　　　　　　　　　98009344

E-mail 系列：7

商務英文 E-mail

編　　著◎張瑜凌

出 版 者◎雅典文化事業有限公司

登 記 證◎局版北市業字第五七〇號

發 行 人◎黃玉雲

執行編輯◎張瑜凌

編 輯 部◎221 台北縣汐止市大同路三段 194-1 號 9 樓

電　　話◎02-86473663　傳真◎02-86473660

郵　　撥◎18965580 雅典文化事業有限公司

法律顧問◎永信法律事務所　林永頌律師

總 經 銷◎永續圖書有限公司

　　　　　221 台北縣汐止市大同路三段 194-1 號 9 樓

　　　　　EmailAdd: yungjiuh@ms45.hinet.net

　　　　　網站◎ www.foreverbooks.com.tw

　　　　　郵撥◎ 18669219

　　　　　電話◎ 02-86473663　傳真◎ 02-86473660

　　　　　ISBN：978-986-7041-83-8

初　　版◎2009 年 08 月

定　　價◎ NT$ 220 元

雅典文化 讀者回函卡

謝謝您購買這本書。
為加強對讀者的服務，請您詳細填寫本卡，寄回雅典文化；並請務必留下您的E-mail帳號，我們會主動將最近 "好康" 的促銷活動告訴您，保證值回票價。

書　　名：商務英文 E-mail
購買書店：＿＿＿＿＿市／縣＿＿＿＿＿＿＿＿書店
姓　　名：＿＿＿＿＿＿＿＿　生　　日：＿＿年＿＿月＿＿日
身分證字號：＿＿＿＿＿＿＿＿＿＿＿＿＿＿＿＿
電　　話：(私)＿＿＿＿＿(公)＿＿＿＿＿(手機)＿＿＿＿＿＿
地　　址：□□□＿＿＿＿＿＿＿＿＿＿＿＿＿＿
E - mail：＿＿＿＿＿＿＿＿＿＿＿＿＿＿＿＿
年　　齡：□20歲以下　　□21歲～30歲　□31歲～40歲
　　　　　□41歲～50歲　□51歲以上
性　　別：□男　□女　　婚姻：□單身　□已婚
職　　業：□學生　　□大眾傳播　□自由業　□資訊業
　　　　　□金融業　□銷售業　　□服務業　□教職
　　　　　□軍警　　□製造業　　□公職　　□其他
教育程度：□高中以下（含高中）□大專　□研究所以上
職 位 別：□負責人　□高階主管　□中級主管
　　　　　□一般職員□專業人員
職 務 別：□管理　　□行銷　　□創意　　□人事、行政
　　　　　□財務、法務　　□生產　□工程　□其他＿＿＿＿
您從何得知本書消息？
　　　□逛書店　　□報紙廣告　□親友介紹
　　　□出版書訊　□廣告信函　□廣播節目
　　　□電視節目　□銷售人員推薦
　　　□其他＿＿＿＿＿＿＿＿＿＿
您通常以何種方式購書？
　　　□逛書店　　□劃撥郵購　□電話訂購　□傳真訂購　□信用卡
　　　□團體訂購　□網路書店　□其他＿＿＿＿＿＿＿＿
看完本書後，您喜歡本書的理由？
　　　□內容符合期待　□文筆流暢　□具實用性　□插圖生動
　　　□版面、字體安排適當　　□內容充實
　　　□其他＿＿＿＿＿＿＿＿＿＿
看完本書後，您不喜歡本書的理由？
　　　□內容不符合期待　□文筆欠佳　　□內容平平
　　　□版面、圖片、字體不適合閱讀　□觀念保守
　　　□其他＿＿＿＿＿＿＿＿＿＿
您的建議：
＿＿＿＿＿＿＿＿＿＿＿＿＿＿＿＿＿＿＿＿＿＿
＿＿＿＿＿＿＿＿＿＿＿＿＿＿＿＿＿＿＿＿＿＿

2 2 1 - 0 3

台北縣汐止市大同路三段 194 號 9 樓之 1

雅典文化事業有限公司

編輯部　收

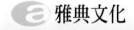

雅典文化

為你開啟知識之殿堂